CW01395012

Yes, there are Alternatives to Plastic
Returning Planet Plastic to Planet Earth

Jane Stewart Adams

Grosvenor House
Publishing Limited

This book is published by
Grosvenor House Publishing Ltd
Link House
140 The Broadway, Tolworth, Surrey, KT6 7HT.
www.grosvenorhousepublishing.co.uk

A CIP record for this book
is available from the British Library

ISBN 978-1-83975-766-2

For Mother Earth

'Cherish the natural world, because you're part of it and depend on it.' – Sir David Attenborough

Table of Contents

Introduction		vii
Chapter one	Memories of life before plastic	1
Chapter two	Plastic, a global problem	20
Chapter three	Plastic inside us	29
Chapter four	Plastic free?	39
Chapter five	Land matters	47
Chapter six	Learning from the past	53
Chapter seven	General advice	61
Chapter eight	Consigning petrochemical plastic to history	74
Chapter nine	What should we recycle?	83
Chapter ten	Let's end single use plastic!	93
Swaps	Thinking Planet first, not plastic	98
Last thoughts – what a plastic-free world would look like		274
Games to raise awareness		294
Setting up an Alternatives to Plastic Demo Table		298
Recommended Organisations/Websites and Books		307

Introduction

The dictionary definition of plastic is anything that can be shaped/formed or is pliable. Therefore, materials can be 'plastic' without being made of the petrochemical plastic that is destroying our precious planet. It's ironic but modern plastic, starting about two hundred years ago, was invented to preserve the world's natural resources. We now need to end petrochemical plastic before it ends all life on Earth!

Surprisingly, according to the British Plastics Federation, horn and tortoiseshell was used as the first plastics in 1284 by The Horner's Company of London. There was no further development until the 1820s when cellulose, plant-material, and vulcanised rubber, made from the Rubber tree came in. A natural latex gum named Gutta Percha, produced from the sap of the Palaquium gutta tree, and Parkesine, synthetic ivory, replacing elephant-ivory for billiard balls, started to appear at the same time. A telegraph cable made of Gutta Percha was laid under the English Channel between Dover and Calais in 1852. Co-incidentally, and much to my surprise, I discovered that one of my ancestors was a Gutta Percha Boot Moulder in the early 1900s.

Tortoiseshell, which nearly wiped out the world's tortoise and turtle populations was used to make combs, spectacle-frames and boxes, among other luxury items. It was mainly made from the shell of the still highly endangered, and often poached, Hawksbill Turtle that nests on beaches in Brazil. Their hard shells, that are uniquely and beautifully patterned, are given a high polish and still used today for many top end items and jewellery. It is a wonderful shell that only the turtles need!

Then came Bakelite – Polyoxybenzylmethylenglycolanhydride - the first hard synthetic plastic, in 1907. It was used for all sorts of things including casings for the first TVs. I still have a light-pull and a belt-buckle made in dark brown, passed down in the family, probably left in a drawer, and a nifty, still-working, meringue-maker in green. It was a fragile material and shattered if dropped.

It's utter madness that we are in thrall to the petrochemical industry and Earth-destroying plastic when wonderful natural materials are all around us and should be used once again. The tinkering of natural bioplastics by a multitude of scientists over decades has resulted in the virtually indestructible, highly toxic substances we have today. The timeline for the development of the plastics industry is extensive. I don't want to bang on about its birth when I know so many of us are now looking forward to the demise of petrochemical plastic. If you're interested in seeing how it all started, the full timeline can be found on the British Plastics Federation website.

Researchers at the University of Newcastle, Australia found that just opening a plastic bottle and even unwrapping a chocolate bar releases tiny particles of plastic into the air which we're breathing in.

No-one wants plastic in their lungs! We can't compromise the health of the world any more by wrapping every item under the Sun in plastic. We must avoid it whenever we can until safe natural alternatives are found and used commercially.

It's going to be a long uphill struggle as even plastic is wrapped in plastic!

Lovely/well made, natural and sustainable are the 3 important things we should be looking for in alternatives to plastic.

I'm just as much to blame for Planet Plastic as anyone else! There, I've said it. Now let's move on. Judgement never allows for growth. We will all have to pull together to heal our planet. This is a world crisis and it's no good playing the blame game. We all caused it. Let's all solve it. Check egos at the door! 'As long as I and my family are all right' is the attitude that never helps anyone. We are **ONE** on this planet and what we do or don't do for each other, matters big time.

- **No**, I don't have a scientific background, nor am I an expert on plastics, but I do daily research, gathering information from multiple sources. As a concerned world citizen, I'm trying to make sense of an extremely complicated subject at an exceptionally difficult time. I apologise for any unintentional errors.

- **No,** I'm not 100% plastic free. I don't think anyone can be at this point. I just take every opportunity to cut out plastic on my budget. I also live chemical free. It can be done, even on a tiny income. We don't need synthetic chemicals, sprays or potions in our households because there are so many natural alternatives that have been effectively used for centuries. Good fresh air is better than room/fabric freshener any day.

- **No,** I don't live a zero-waste lifestyle, but wish I could.

- **No,** I don't have any ego about what I'm hoping to achieve. The only thing I pride myself on is that I'm a kind, caring and decent human being. This is for our glorious Mother Earth, our one and only Home-planet.

- **Yes,** I grew up when there was very little plastic.

- **Yes,** I want to tell a human story. I'm trying to see the whole distressing picture. I want to make it clear how serious the situation is. (I have spent a lot of time in tears doing the extensive research needed for this book.) I want to make it clear, though, that we can **all** do something to help locally and globally.

- **Yes,** I'm learning as much as I can to educate myself and others.

- **Yes,** I want us to question our lives of convenience.

- **Yes,** I avoid anything made of artificial fibre if I possibly can.

- **Yes,** I hate waste and try to avoid it whenever possible by reusing, recycling, repurposing what I can and composting food scraps at home. I also have a worm bin/wormery.

- **Yes,** I save water and use grey water whenever possible.

- **Yes,** I plant for life and help wildlife thrive.

- **Yes,** I'm doing this for Earth, our spectacular, wonderous, awesome blue jewel of a planet, with all Her fabulous lifeforms. That includes us!

'Unless someone like you cares a whole awful lot, nothing is going to get better. It's not.' From *The Lorax*, written by Theodor Seuss Geisel, known as Dr. Seuss.

Chapter one

Memories of life before plastic

Before plastic took over, our lives certainly didn't revolve around household waste management the way they do now if we love the planet. The world was full of hope after the unspeakable losses of World War II. My dear parents told me that life was unbearably dreary at home during the war. There was bad news daily, deadly air raids, bad or no food because of rationing, terrible dull clothing, basic crockery with no decorations, with blackout curtains and dreary colours wherever you turned. You were constantly reminded of the dangers with the air-raid wardens shouting, 'Put that light out!' if only a chink of light could be seen from any window at night. Everyone was also 'digging for Britain' having turned any tiny piece of soil into a veg garden to feed a starving nation.

One of my father's jobs in the Royal Navy was to help protect food supplies coming from other countries in the Atlantic convoys stalked by German U-boats. He survived the sinking of one of the destroyers he served on but lost many friends and colleagues and everything he had onboard including all his clothing. He was picked up by an American aircraft-carrier and given a complete American flyer's uniform, including the iconic leather flying-jacket. Mum had quite a shock when he turned up wearing this at her door on leave, but I bet he looked handsome!

In the decade after the war, thousands of prams full of smiling babies appeared on the streets after women were asked to repopulate the country. I was one of those little souls that was meant to herald a kinder brighter future. Sadly, we've been too easily led by those whose intention was to make money and gain power to the detriment of our once thriving planet.

One of my enduring memories in the early Sixties is playing on the bomb-sites, where the massive bombs of WWII had fallen, that were awaiting development in some towns. They were full of hidey-holes and smothered in lovely wildflowers that needed upheaval to germinate. We didn't give a thought for the memorial of devastation they represented. To our fledgling minds, the rubble-strewn remains of lives long gone were the pure fun of pirate ships and castles. I clearly recall telling my mother, in local parlance, that I was 'going up the bomb-site.' This would horrify any modern mother but that is how it was at this time. We were trusted not to do anything too daft. I don't remember being injured from playing on any of these sites. We did have to avoid the occasional rusty nail. Most of the windows and doors had thankfully been repurposed, but there were very few broken bottles or cans and, of course, no horrible plastic that is everywhere now. Any rubbish on these sites, like paper or cardboard, was organic and did no lasting harm to the environment, and was a cosy home for insects while it safely rotted down.

In the home

Most homes still had coal-fires for heating when I was a child. These proved a mixed blessing because of the stench and dense smog that hung around the towns. I remember being sent out into the back garden armed with a coal-scuttle/bucket to shovel coal from our brick coal-bunker when I was a bit older and stronger. Carrying out this task on freezing cold snowy Winter days was exceptionally unpleasant but that is what most of us had to do before gas-fires, and, later, central heating.

It may be hard to imagine now in an era of central heating, every chimney in every street spewing black smoke. I remember wrapping my wool scarf around my nose and mouth to give myself a little protection from the choking coal-dust in the air in cold weather. We burnt most of our rubbish in our homes, further adding to the smell, including sanitary/period wear which was cotton and paper. All cardboard, paper and greasy packets also went up in flames. Sometimes you could see little pieces of burnt paper flying out of the chimneys, a potential fire-risk.

We had Pig-bins on the streets in my town, for the one and only local farm, where we deposited any food scraps. As we shopped in season, and cleaned our plates at mealtimes, there was very little waste. If I refused to eat my lunch my mum threatened to serve it cold for tea because we really couldn't afford to waste anything. I'm pleased to say it never happened. Joking aside, we were well fed on a meagre budget in the austere 1950s. Food-rationing had only just finally ended in 1954, the year I was born. The nation had suffered this for 14 gruelling long years.

One of my household chores when I was older, most children were expected to do them, was to scrape the plates into the Pig-bin 3 doors down by the kerb. It was admittedly a ghastly smelly job but I don't remember our aluminium dustbins ever overflowing with stinking rubbish, and they were much smaller than modern household wheelie-bins. We had no trouble with vermin or birds as the bins were secure if the lid was on firmly. They were called 'Dustbins' originally because the leftover ashes/dust from the fires went into them, safely. The Pig-bins were collected daily in the evening and replaced in the early morning, so the pong didn't hang around long!

Our local schools had them as well. We even had 'Pig-bin Monitors/Plate Scrapers' in my secondary school. Guess who got the job?! I clearly remember standing over the bin after I'd had my lunch. As a prefect, my task was not only to make sure that every last precious morsel was scraped from the plates into a large metal bin but that none of the washable stainless-steel cutlery landed in the pigs' dinner!

Towns

On every high street in those days the Greengrocers were full of lovely fresh produce kept in freezing cold open-fronted shops and doled out into customers' wicker baskets, nets or cotton bags. We could buy exactly the number of items required, and weren't forced to buy a pack of anything. Again, no waste.

The bakers didn't use anything but paper and cardboard for packaging their delicious confections, free of additives and horrible soya flour. In the delicatessens, cheese was cut off the block on a marble slab, using a cheese-wire, with clean un-gloved hands, regularly washed at the shop sink. You could also request the exact amount of cold meat you wanted, that was delicately sliced by the fascinating meat-slicer machine. It was then picked up in a square of greaseproof paper and securely wrapped up in a large sheet of white butcher's paper, which of course the butchers themselves used.

The butchers' shops were spotlessly clean, often with sawdust on the floor for any drips, and meat was kept on large aluminium trays in the chiller displays, with nothing wrapped in plastic. You could also buy eggs and collect them in your own repurposed cardboard egg-boxes. I'm pleased this is still done in the UK.

Delicious sweets, with no artificial additives, if they weren't from the many sweetshops or the 'Pick 'n' Mix' in Woolworth's and put into white paper bags, all came wrapped in paper, were in cardboard tubes, or paper and foil. Crisps/chips, biscuits/cookies and snacks also came in waxed paper and/or cardboard. Sadly, all types of packaging ended in landfill if it wasn't burnt, as there was no household recycling at that time. At least most of it was organic, biodegradable and did little or no damage to the soil, and in some cases improved it.

Nothing tastes as good as it did, with palm oil fat and soya bean flour, very little salt, hardly any real sugar and stuffed with artificial everything else! The modern-day food industry is doing a splendid job of poisoning us, and adding plastic to the mix!

The old-fashioned Ironmongers, obliterated by modern DIY super-chains, were wonderful shops. In these smaller but well-stocked stores you could buy anything needed for the household. There were also rows upon rows of boxes stacked on shelves full of solid iron nuts, bolts, screws and nails, etc., of every possible dimension, not like a lot of the soft, cheap nasty Aluminium ones around now. These were counted out like heavy sweets into strong brown paper bags or boxes as there were very few packs of anything. If you required 6 nails for a job, then that is what you bought plus a couple of spares.

Every tool and household requirement imaginable was available on request. I can remember seeing all the metal buckets and dustpans hanging up on a piece of strong string outside the door. Most of these products were made in a Britain that still had a manufacturing heritage but some started to creep in from Japan, and were partly made of or in plastic. I remember the first sets of tools that were in the new plastic blister-packs or in a plastic pouch instead of a leather one. You could still buy things in plant-based Cellophane in those days but plastic was coming in fast. It was a cheaper option for retailers, and enticing for cash-strapped customers.

I clearly recall my father placing some obscure household item on the glass counter, asking: 'Have you got one of these?' and the brown-duster-coated male assistant scratching his head before climbing the sliding wooden ladder to root around in one of the many dusty carboard boxes on the uppermost shelf. Then returning with a triumphant smile, he plonked the 'treasure' in front of us with a brisk: 'Will this do?'

It really was like the 'Four Candles' sketch from *The Two Ronnies*, so fondly remembered by fans all over the world. I loved visiting these dark and mysterious emporiums with my dear father on our Saturday mooch around the shops in my home town, and often Lewisham, where we had lunch in the wonderful old-fashioned department store Cheeseman's. It was always a fun day out with my father, who was often too sick to go out and about much when I was growing up. He was a complicated man, a bit grumpy, older than my friends' dads, and damaged by a bad war, but these days for me were golden. I learnt so much on these trips that most girls in the 1960s didn't know, that has put me in good stead as a householder.

The Haberdashery, Drapers and Wool shops, so rare now, were a delight for my dear mother, who knitted and sewed nearly all the family clothes even though she'd had a high-powered professional career from the age of 40. On the rare shopping-trips with her, I loved the bright wools, and materials in huge bolts on the shelves, and looking at the reels of silk and cotton thread in such vivid colours. No wonder I became an Artist!

Now most materials for clothes-making are made of Nylon and Polyester.

Milk was delivered daily to our doorsteps in returnable glass bottles and if you were lucky your friendly 'Milky', who kept an eye on his elderly and vulnerable customers, could also supply eggs, cheese, yogurt, and soft drinks in returnable glass bottles. Most of us were woken up each morning by the clink of bottles on the electric milk-float and the cheerful whistle of the milkman coming up the path.

The Sixties and plastic

Plastic toys were just starting to creep in from Japan in the 'Swinging Sixties', usually financed by Post World War II USA. I remember two of my novelty toys from that period: a plastic-foam 'Fred Gonk', a quirky British character with a flat cap and braces from the film-animation and song '*Right said Fred*', sung by the wonderful Bernard Cribbins, and my beloved 'Jenny' donkey made of the same material. Within a few years both had disintegrated into toxic orange dust. Even then I knew it wasn't good to breathe that in, and had to carefully bin both of them. I dread to think what this sort of toy has done to the environment. We weren't aware of the potential planet-polluters we had in our homes, even our beloved dolls/action figures and Teddy-bears, in these early days of plastic and chemicals.

Trendy plastic handbags and matching shoes started to appear in the early Sixties. Then plastic coats became the 'in thing'. I had a lovely red 'midi' one that after years of use 'peeled away' and had to be thrown out. I also had some shoes made of the same strange material. There were also plastic brushes, combs, picnic-items and cleaning equipment like mops, brooms, dustpans and brushes to name but a few new products rapidly filling supermarket shelves. No food was smothered or entombed in plastic at that point. That happened a couple of decades later. By the late 1980s there was no getting away from it!

I do think we were kept in the dark by the powerful plastic industries. I'm sure they could see what a plastic world would do to the ecology of the future but just kept extruding it anyway. It was an absolutely atrocious lack of forethought on their behalf and took many intrepid explorers to expose the damage plastic has done to the planet for decades. Thank the powers that be that they did!

The plastic carrier bag had just started to emerge by the middle Sixties but you didn't see many. We had our own various types of bags to use. Plastic milk bottles hadn't been considered until a Polyethylene Terephthalate or a PET plastic bottle was patented in 1973 by a Dupont scientist. It was only when supermarkets started popping up that these plastic milk-containers appeared alongside the glass bottles, for 'convenience'. No-one liked them at first until they realised that they were cheaper than having a daily delivery. That, sadly, was 'the death of the milkman'. Local dairies, most towns had one, closed after that. Not surprisingly, door-to-door is rapidly returning because of the fear of pollution, mainly from supermarkets, those vast cathedrals to plastic.

Of course, there were no revolting used plastic nappies/diapers anywhere when I was a child. (Surprisingly, a nun invented plastic ones to make baby care easier for her young mothers.) The only choice was Terry Cotton Towelling and a toilet in which to dispose of the unmentionable. There were certainly no plastic dog-poop bags 'decorating' the trees, or in Africa bags they call 'Flying toilets', I'm sure I don't need to go into detail. Plastic bottles of pee left in lorry/truck laybys, and in our coastal towns with busy Continental traffic, were a nightmare yet to happen. The birth of plastic bottles, and thinner materials, coupled with a 'So what?!' attitude has been the cause of these daily horrors that now plague the entire planet.

One of the worst apocalyptic stories now is the pollution of the lifeblood of Egypt, the incredible Nile river-system that is being poisoned and suffocated by plastic waste daily. In Cairo, there are piles of waste on rooftops everywhere and the fish locals depend on are contaminated with plastic. This situation needs urgent attention or the mighty life-giving river will die. It's no exaggeration to say that if this isn't addressed, the people and wildlife will starve because of plastic pollution!

Synthetics at home

Nylon, a plastic material that could also be spun into fine fibres, invented in New York and London by Dupont in 1938, started to replace a lot of natural fabrics. There were the Bri-Nylon shirts for men that could be hung over the bath to drip dry, and Nylon stockings, first demonstrated at the New York State Fair in 1939. Later in the middle Sixties Nylon tights came in. Stockings used to be expensive because they were made of Silk. Nylon was far more affordable and stretched to fit any shape.

I remember taking my first pair of tights gleefully from their cardboard packet and wondering how something so tiny was going to fit a growing girl! The first tights were a bit of an odd shape, straight up and down. They were certainly a struggle to get into, and generally uncomfortable, but I was glad not to be wearing an even more uncomfortable suspender-belt and stockings that didn't feel good, or seem decent, with a miniskirt!

Crimplene, developed in 1959 in the UK, didn't crease and kept its shape when used for clothing. It was the 'in thing' for the iconic Sixties fashion boom. I had several dresses made of the fabric as well as the usual cotton ones. About the same time, cheap Nylon ladies' headscarves came in and seemed to be everywhere in every colour, replacing cotton or silk. Anything made of Nylon creates static electricity. I can recall 'crackling' clothes and hair standing on end!

I also remember the first synthetic shirts looked shiny and felt 'plasticky'. We all thought these various artificial fabrics were wonderful: easy washing and no ironing. There was a lot of experimenting going on. We had no idea that clothing made of plastic fibre would become the huge global crisis it is now. Only the makers knew what chemicals these 'modern' clothes were actually made from. It didn't seem to matter to anyone else at the time. We turned a blind eye even though some clothing had a strange smell and could irritate the skin. How wrong we were! Later Polycotton became the new mixed material of choice and looked and felt more natural even though it's now as much of a problem as anything fully synthetic.

Life was hard and far from perfect

I certainly don't look at those times through rose-coloured specs. I'm ashamed to say we were too easily seduced by plastic, but it was a better world of well-made and hard-wearing, with no built-in obsolescence. Unfortunately, the downside was that most people smoked everywhere. I spent a great deal of time with streaming eyes and a tickly throat that made me cough. I clearly remember not being able to see the screen at the local cinema, because of the sickly pall of smoke around me, and feeling ill on public transport. The stench of coal, the dirtiest fossil fuel, and cigarette smoke filled the air wherever you went. Even doctors smoked during consultations!

Clothes, hair and curtains reeked of both and what should have been white ceilings were often tinged yellow. Even at home I wasn't saved from it because my father smoked like a chimney and stank like an ashtray. In Winter we had to stay warm by the living room fire, having to put up with him chain-smoking in what mum called 'his dirty corner'. He had nicotine-stained fingers and beard. I think most of his wages went up in smoke.

Cigarette smoke contains a lethal cocktail of over 7,000 chemical compounds. These include arsenic, formaldehyde, hydrogen cyanide, lead, nicotine, carbon monoxide, acrolein, and other poisonous substances. Over 70 of these have been proved carcinogenic, causing cancers. Why would anyone want to willing poison themselves?!

Both my dear parents passed from smoking-related diseases even though my mother hadn't smoked for years, everyone smoked in the services during WWII. My mother probably developed heart disease and finally cancer from passive smoking, that for decades she was unable to get away from in our home or at work, where she shared an office with another chain-smoker. My father passed horribly of bronchial pneumonia because his lungs were so congested even though he had finally stopped smoking 10 years earlier. All I can say is thank God for the smoking bans in the UK. No-one from younger generations can even imagine how bad it was coupled with coal-smoke everywhere. We should be thankful that with these necessary bans we can all breathe a little better in public spaces.

In those days, cigarettes came in metal foil with a paper backing in cardboard boxes with a flip lid. Later on, they were in a Cellophane wrapper too, which was at least biodegradable. Now they are packaged in plastic and have plastic filters. Some of these are made of recycled plastic but that's no better. There is far less smoking in the UK now but Worldwide cigarettes and their packaging are one of the major polluters. Mindlessly flicked fag-ends/butts find their way into global river-systems and oceans every second to poison and kill wildlife.

Cape Town in South Africa, sickened by the toxic unsightly pollution on its beaches, placed discarded telephone-poles cut to size and painted to resemble giant cigarette butts in prominent spots. Their aim was to discourage cigarette-smoking and raise awareness of carelessly-discarded cigarette-ends. The initiative was called 'Kiickbutt' in Africans.

I love this idea. Maybe I should make something similar to discourage the regular 'flickers' who queue in the early morning for the Post Office/General store next door, along with the sign: **'I, and the planet, don't want your butts!'**

The good news in the USA is that legislators in some states are hoping to bring in bans. These are led by the, in my humble opinion, more enlightened state of California that is thinking of banning products with single-use filters made of cellulose acetate. This could force cigarette companies to take responsibility for the environmental impact of all their products. It would be a huge step towards a cleaner and safer world for all life, and maybe result in fewer smokers.

There are cigarette-butt recycling schemes in Canada but I don't see recycling any type of plastic as the ultimate solution to the world crisis. There are so many types that can't even be recycled. Recycling plastic is just downcycling and delaying inevitable global pollution.

Money matters

Growing up in the UK, our money was mostly paper and not the majority of plastic it is now with debit/credit cards and plastic notes that are meant to last for decades. They will, but are slippery, don't easily stay in purse or wallet, jump out of tills and will make the plastics industry happy for a long time to come. I was horrified to see these were the solution to 'tatty' paper notes that didn't last. Plastic notes could create complications down the line when they need to be incinerated. Maybe the idea is to recover the energy but that will not help slow down plastic production, and might even ramp it up again.

Food in plastic

I can hardly believe that the plastic bottle/container/food packaging nightmare actually started when I was still at school. The USA embraced it long before the UK and it wasn't so obvious here. It sneakily crept up on us. Before we even realised it, the whole world was plastic.

I worked in a supermarket after school on Thursday and Friday evenings and all-day Saturday and have memories of the different plastics coming in. There was still a lot of real Cellophane, made of plant fibre, and cardboard packaging as well as the usual tins and jars. However, tins were not coated inside with a thin layer of plastic, as many are now, to prevent corrosion. The shelf-life of foods wasn't long. Everything was used up far more quickly. There was no such thing as 'Best before'. Common sense was regularly used along with the Eyeball, Sniff and Quick Fingertip Taste Test.

A 'Best before' date is only a quality guide. Use your own judgement with the aforementioned test to see if the product is still good to eat. Many foods are still good months or even years later.

'Use by' date is the safety mark there to protect us, and should never be ignored. Some foods do expire and go bad, causing sickness, and should not be eaten after this date. Peanut butter, because it contains oil that will spoil, is a good example.

Plastic tubs

We first saw the container/tub type of plastic food-packaging in the UK after the home fridge-freezer became more common. (Ice-boxes with blocks of ice delivered by the 'Iceman' were used before refrigeration.) Plastic pouches appeared in the new freezer shops from the late 1960s for frozen fruit and veg, with the first plastic tubs being developed for margarine, ice-cream and yogurt. Before that ice-cream was only doled out from a metal freezer, or extruded from a tap, from ice-cream vans that plied the streets daily, or in sweetshops with a chest freezer. You could often buy a block in carboard from the same shops. Plastic free, cardboard-packaged blocks are still available from some UK supermarkets. Most are Vanilla and used for wafers. For variety you can add something fruity or chocolatey to the bowl to avoid buying fancy ice creams in plastic tubs. Fortunately, once washed and dried cardboard and carboard tubs can be recycled as long as there is no plastic lining. Check recycling details on the products first.

It's about time ice cream producers cut out these ubiquitous plastic boxes that we've collected over decades and have piles of sitting in cupboards and drawers. Any plastic tub can be reused or recycled but that is not the way we cut out plastic. Avoiding it as much as possible is the **only** answer until natural and sustainable alternatives are commercially viable.

Schooldays

Most schools when I was growing up served a proper meal with water at lunch time. If they didn't have a kitchen onsite, food was transported to the school in large aluminium containers with lockable lids. If children didn't have the school lunch, they took a sandwich in a tin or reused brown paper bag, with an apple, orange or banana. Sometimes there was the delicious treat of a Lyon's Individual Fruit Pie in its cardboard box.

We didn't need extra drinks as there were school fountains to quench our thirst. I found my way to a wall-fountain whenever I asked to go to the toilet. They were a bit busy during breaks but nobody needed to carry a water-bottle. That everyday habit originated in the USA from hotter states where regular hydration is essential. In the early days pop bottles were glass, and paper straws were used. Plastic straws turned up after WWII. Some drinks companies that used glass bottles here in the UK had a deposit return scheme that boosted many a child's pocket-money. I believe it was the same in the USA.

We must return to this sort of scheme if we're to clear up our filthy streets littered with cans and plastic bottles. Many countries, like Norway, Italy, Germany and Japan, have successfully introduced machines for this purpose in airports, stations and other public places. Although I don't believe they will reduce plastic production, they will prevent littering with the incentive of the small reward for their return.

Street-food except for fish and chips in newspaper, ice cream in a cone, and coffee and tea at a stall in ceramic cups, was unheard of when I was a child. No-one walked around with a box of food, drinks-bottle or plastic-lined coffee cup in their hand because there were no such things. Eating in the street was heavily discouraged by most parents, and should be once more to keep our towns and cities clean. Hand-washing when having been out and about was also the norm. My dear Mum's firm instruction to 'Go and wash your hands!' rings in my head to this day.

Life was far from perfect but it was simpler. Something we must return to if we want to save the world from our horrendous consumerist waste. The responsibility lies with each person on Earth. If we want a healthier, safer planet for all living beings, we can't let our arrogance and cavalier attitudes to world problems rule us anymore.

As Albert Einstein said: *'The only thing more dangerous than ignorance is arrogance.'*

We can return Planet Plastic to Planet Earth if we all pull together. The habit of replacing something because it's old or out of fashion or because we need some 'Retail therapy' has to end. Our children must be taught to value all things and not constantly look for something 'better' or the latest trends that give them momentary pleasure.

'Keeping up with the Joneses' with a new kitchen or bathroom, or replacing things in the house because they are the wrong colour or style, must end too. We can always paint, refurbish, replace cupboard doors and worktops in the kitchen instead of ripping the whole lot out and putting in an entirely new one. Bathrooms can also be upgraded without getting rid of everything.

We can't subject our beleaguered planet to any more unsightly and toxic waste that is endangering and killing life across the globe.

I put anything still good but unwanted outside my house with a 'Free' note on it and it's usually gone within minutes!

You are very lucky if you live in an up-and-coming area that is becoming more eco-friendly as it evolves but the truth of life in the UK is more complicated at present.

- Most of us don't have a farmer's market or zero waste/plastic free business nearby.
- Some of us don't have the wherewithal for an organic and plastic free fruit and veg service.
- A lot of us reside in small towns with rapidly-disappearing local shops.
- Some of us don't drive and depend on infrequent local transport.

So, life for the plastic-free warrior can be hard without online shopping, but we can all do something today.

Reducing our reliance on fossil fuels, petrochemicals, and avoiding plastic wherever possible, are the first steps to healing Mother Earth.

All life is precious and vital to a healthy, diverse and abundant planet.

We must fight to protect Her, whatever the cost!

Chapter two

Plastic, a global problem

'There is absolutely no logic in wrapping something as fleeting as food in something as indestructible as plastic. Plastic food and drink packaging remains useful for a matter of days yet remains a destructive presence on the Earth for centuries afterwards.' - Sian Sutherland, Co-founder of A Plastic Planet

One Australian school experimented with removing its trash cans in 2019. The students had to take home any rubbish they made and dispose of it responsibly. This is a great idea. Once you see your mostly plastic rubbish, and have to carry it around all day, habits will rapidly change. We must also avoid buying anything that has the word DISPOSABLE on it and investigate the end of life of anything that doesn't display vital information on how to correctly recycle or ultimately safely dispose of.

The refill revolution

Refilling is becoming the 'new thing', after a brief resurgence in the 1970s, when you risked falling into large bins of flour and cereal! This method of dispensing goods is mostly found in zero waste and plastic free highstreet shops and even in some more enlightened supermarkets. Thankfully, this has been upgraded for the new century with strict hygiene rules and regulations that didn't seem to be present in the past. Refilling from large plastic hoppers and bulky containers is not ideal but it is a start and saves on single use plastic overall.

In 2020, one of the large UK supermarkets teamed up with well-known brands to offer refill points at one of its stores. The trial proved popular with customers. I'm not certain this will work in every supermarket once it's scaled up unless we have whole supermarkets dedicated to refilling.

We may need this way to shop for cereals, nuts and pulses, etc. Fortunately, flour and oats still come in paper. As I mentioned, in the past Cellophane was common but lost out to plastic. If we want the same sort of convenient see-through packets, we must go down the plant-material route again. Personally, I would like to see a total rethink on **all** packaging, see-through or not.

I believe shopping in general is going through a complete overhaul. Our highstreets will soon look very different. There will be far fewer chain-stores/retail outlets, because I believe their day is over. Space will be used more effectively for local communities and projects to help improve the environment. I envision more wildlife-friendly and greener towns full of independent plastic free and zero waste shops, tool, toy and tech libraries alongside repair and clothes swap shops.

The worldwide Right to Repair movement is gathering pace. The good news for the UK is that this right to repair was made law in the Summer of 2021. Manufacturers are now legally obliged to make products repairable and to provide spare parts. The overall aim is to extend their life by up to 10 years. They must also be energy efficient, saving money and reducing carbon emissions.

In the past you would have expected well-made and repairable products to last a lot longer than the proposed 10 years, and most did. Nevertheless, this is a positive and very welcome step forward to reduce e-waste mountains that are littering and poisoning the planet.

Ideally, zero waste is what we need:

'The conservation of all resources by means of responsible production, consumption, reuse, and recovery of products, packaging, and materials without burning, and with no discharges to land, water, or air that threaten the environment or human health.' – Zero Waste International Alliance

The 5 Pillars of Zero Waste (I have paraphrased a little here.)

1. Visit landfill and incinerator sites to see what is being wasted, and make a plan for Zero waste in the future.
2. Make producers accountable so they design out waste from their businesses. *(I would add: Especially single use plastic.)*
3. Encourage consumption that respects ecological limitations.
4. Build infrastructures to ensure that all materials are reused, repaired, recycled, or composted. *(Not happy about plastics being recycled and kept in the*

loop. Ocean plastic to plastic again is also not a good idea as plastic that has been in a watery environment may harbour dangerous bacteria and will need stringent cleaning.)

5. Prioritize social and environmental justice so that everyone's voice, especially those workers and communities on the frontline of the waste crisis, is heard and respected.
   (I would add a 6th:
   Wealthy countries should help less developed ones and no waste should be shipped from them for 'disposal'.)

Each country should have a duty to clean up its own backyard and not spread the waste around the world for others to clean up in an out-of-sight-out-of-mind mentality. Rainy/Monsoon conditions regularly wash waste from open landfill into rivers and oceans literally creating tsunamis of plastic and general filth that also carry disease.

Recent Geotagging has revealed that plastic bottles can travel thousands of miles along river-systems to seas and oceans. Invasive species can also hitch a ride on ocean plastic/litter. It's known as 'Plastic rafting' and is another global problem. With the rivers too contaminated for drinking or irrigating crops, communities that depend on these waters are doomed to scarcity and ultimate poverty. These are heart-breaking scenarios we shouldn't be witnessing in the 21st Century!

Plastic bottle recycling schemes

As far as I can see this sort of recycling that is becoming more worldwide will just encourage people to buy plastic instead of cutting it out altogether. At the very least they won't be thinking glass or aluminium. It will be a short-term solution to cities' plastic problems but in the long term the plastic is still there.

The aim for our health and that of our environment should be to avoid plastic products altogether and use natural sustainable alternatives. I really don't understand the throwaway mentality. If something can be used again, or given a new life, it should be. After WWII even one small piece of string was precious. I clearly remember my parents and grandparents rolling up parcel-string and putting it safely in a drawer for further use. I still do it myself.

Plastic waste is all around us

Messy gardens like my own mini jungle are wildlife havens but the amount of plastic I find here is worrying because I live in a windy coastal town, and also have dirty neighbours who carelessly throw things onto it. When I'm in the back garden, I find blown-in polystyrene packing peanuts from all the eateries around me, bits of old greying sweet-wrappers and crisp-packets, bottle-lids, and plastic film that has embedded itself into the soil over decades. This is one of the 'hidden' problems of plastic pollution that not many people understand, and need to. My front garden is even worse as it is also full of the aforementioned mindlessly-flicked cigarette-ends and cigarette packaging and film. Each time I empty one of my planters at the front because something in it needs a bigger one, I find historic bits of plastic of every description. This upsets me terribly as I never use anything harmful in my gardens. All wildlife is cherished and most welcome to share my little bit of borrowed Earth.

My old house which was once a shop itself is on a main road lined with shops and more eateries than required by any town. Infuriatingly, there are often drinks-cans, bottles and pizza-boxes mindlessly tossed into my planters by the fence. Some are left on top of my two wheelie-bins because the lazy lumps can't be bothered to walk a few yards to a street bin!

I once found a complete bacon sandwich carelessly discarded in my front garden still in its packaging. So, I took it out and put the smelly carton into one of the two street-bins in my stretch of road. It was a rare treat waiting for the local sea gulls. When I came back from shopping, thankfully, it had been eagerly wolfed down.

Mine is one garden out of millions in the UK all suffering in the same way. Most gardeners love and feel a deep connection to our beloved Earth and can't bear to see Her so destroyed. If you do see anything plastic or polluting in your own garden/outside space, no matter how small, please remove it as soon as you can. Dispose of it as well as you are able but never burn any plastic in your own garden or anywhere else. It gives off deadly toxic fumes. (I can't even bear to think of the pollution and danger of melting plastic, as well as the mind-blowing loss of animal life, that has resulted from the Australian infernos of 2019/2020. I sincerely hope that many lessons about the preciousness of life have been learnt from this heart-breaking tragedy.)

Report anything untoward going on in the countryside, like fly-tipping, to the local Council or

Environmental Agency. We must beat this appalling, unsightly and toxic habit! Wherever there is plastic there is environmental damage with the Sun and light photodegrading it, breaking it into smaller pieces that will never go away. (Most gardeners have plastic planters that have shattered after a few years left out in the elements.)

Modern tyres are made of 19% natural rubber, 24% synthetic rubber/plastic, the rest comprising of metal and other components. Small particles of plastic from tyres are in the run-off after rain/flooding, especially on busy roads, and find their way into drains, sewers, watercourses and finally oceans. Once there, they continue to break down into trillions of microplastic particles that do considerable harm to marine environments. Our planet desperately needs Earth-friendly tyres as well as other car-parts that would normally be made of plastic. The car industry must find a way, and has a responsibility to us all, to make tyres eco-friendly as soon as possible. I am pleased that something is being done about tyre-pollution.

One recent innovation is from the *Tyre Collective*, founded by a group of Masters students from Imperial College London, and the Royal College of Art, that attaches to vehicles and vacuums-up tyre-dust before it can enter our atmosphere and lungs. It won a James Dyson award in September 2020.

Another innovation, taken from an older process originally developed by the Soviet Union, is making tyres from Dandelion Rubber. These would be a win all round: good for the bees and the environment.

No place on Earth is safe from plastic pollution

The only way to prevent a worldwide plastic geological layer, that will be discovered in horror by our descendants in thousands of years' time, is to turn off the single use plastic tap now. We must also mobilise to clear up the monstrous amount that is already here suffocating and poisoning our sweet Earth.

It isn't possible to think of plastic in isolation from what is going on worldwide. It has invaded every aspect of our lives and some think we can no longer live without it. We did live without it not so long ago, and we can once more.

It's truly terrifying to know that plastic particles are in the air we breathe. Plastic is shed from everything that contains plastic, especially clothes, bedding and furniture and is contributing to the dust in our homes. It's also falling with rain and snow and being tossed into the atmosphere by the motion of waves. The fresh sea air we think we are breathing in isn't good for us anymore. Particles are also found in the deepest oceanic trenches, and in the life that resides there, and on top of our highest mountains. They've also been found inside glaciers in the Alps and in Antarctic sea-ice. The scientists studying this fast-forming ice believe the local source is fibres from synthetic clothing.

Natural gas is mostly used for heating homes and businesses or for fuelling power plants but it also contains another key ingredient and a building block of plastics, Ethane. Fracking and plastic **do** go together to accelerate climate change. Our scientists don't know what harm plastic particles are doing to us, yet some people still maintain that plastic is not the problem. Oh yes it bloody well is! A massive global one! This should be common knowledge.

Please tell everyone you know and encourage them to cut plastic from their lives. Plastics are destroying the planet. They seriously compromise our health every breath we take. They may even contribute to infertility in humans and many other species. There are already solutions to this crisis, and it starts with each individual doing their best to preserve life on our spectacular and diverse planet.

Chapter three

Plastic inside us

'We need to get angry and understand what is at stake. And then we need to transform that anger into action and to stand together united and just never give up.'
- Greta Thunberg

Greta is talking about climate change here but it can also be applied to plastic pollution. We **do** need to get angry because it **is** destroying life!

A study, published in the journal *Environment International*, carried out on a small selection of healthy women by the San Giovanni Calibita Fatebenefratelli Hospital in Rome, Italy, discovered plastic particles in the placentas of four of the women. Thankfully, the women did deliver healthy babies but the scientists were concerned about the long-term effects of plastic on the future development of a foetus. It could potentially trigger adverse immune responses.

This study should be setting off alarm bells across the globe! Turning off the petrochemical plastic tap at source is the only way we are going to stop this horror-show from escalating!

Plastic is already on our faces with glasses and sunglasses, in our ears with hearing aids and inside us with various implants. When I started wearing glasses at 17, they had heavy expensive glass lenses. Now lenses are cheap lightweight plastic, as are most frames, and you can get 2 pairs of prescription glasses online for less than £20 all in. Even far cheaper reading glasses are sold in supermarkets. New innovations like these are fuelling our throwaway culture but some sunglasses are now being made with Bamboo or wooden frames or entirely out of plant materials. One enterprising maker is even using compressed old jeans for frames but stretch jeans contain Elastane, plastic.

I once bought some rather clever glasses with adjustable lenses and this was on the packet, that I hadn't seen before purchasing:

'Warning: This product contains a chemical known to the State of California to cause birth defects or other reproductive harm.'

Surely, they shouldn't have been manufactured in the first place! As I've never had any success in sending things back to China or of even getting a refund, they're staying in their packet and I'll use them on my Alternatives to Plastic Demo table to illustrate 'really bad plastic'.

Toothpastes have plastic free alternatives in the form of toothpaste tablets/brush mints, pastes and powders and natural toothpastes that come in a metal tube as of old. There are even subscription services available to help us save money as well as the planet. Bamboo toothbrushes can replace the ubiquitous plastic ones. You could try one made of Corn starch if Bamboo is not for you. Look out for brushes with natural bristles and not Nylon. The question is, though, will there ever be an alternative for plastic dentures and bridges that could be harming us? Not everyone can afford expensive ceramic and titanium implants. I sincerely hope something body-safe is found soon. We'd all like a healthier smile!

Hemp, the wonder material of the future?

If the American car-manufacturer Henry Ford could make his 1941 'Bio-plastic Model T' using Hemp in part of its construction, then surely that particular bioplastic could be used for other applications in the future. His revolutionary car could run on Ethanol made from Hemp or any other agricultural waste product used as fuel. His wise words still echo down the years:

'Why use up the forests, which were centuries in the making, and the mines, which required ages to lay down, if we can get the equivalent of forest and mineral products in the annual growth of the Hemp fields?'

Hemp is not only fast-growing but growers often choose to grow it organically, with no pesticides, opting for natural pest-control.

There is now more plastic than we can even imagine!

We didn't have much plastic food-packaging when I was a child. It's crept in since the 1970s and has proliferated worldwide since the 1980s, when ironically the public started to rebel against its over usage. I remember hoarding my plastic bags and washing them out for reuse because I was worried about the amount of plastic I was buying and disposing of.

Shockingly, there are over 5,000 different types of plastic and about 4,000 different chemicals used in food-packaging alone. No scientist actually knows what they might be doing to our bodily systems, those of animal-life or the wider environment. There have been few comprehensive studies to date. Though it was estimated, according to a study carried out by the *World Wildlife Fund*, that we consume a credit-card-sized amount of plastic per week from food packaging and the microfibres surrounding us in the air. In some places these nano-plastics have already entered fresh drinking water. This all adds up to a petrifying plastic hanger's worth in our bodies each year.

Just let that sink in!

'For most of the thousands of chemicals, we have no way to tell whether they are safe or not. And manufacturers may or may not know the ingredients of their products, but even if they know, they are not required to disclose this information.' - Martin Wagner, a biologist at the Norwegian University of Science and Technology and a co-author of a report published in *Environmental Science and Technology*

One of the most over-used plastics is Polyethylene Terephthalate or PET. Patented in 1941, it is not only used for making the aforementioned plastic bottles and food containers/tubs that now litter ever country on the planet, but tape as well. It even appears in traditional solar cells, but that is changing rapidly with a new innovation from the University of Newcastle Centre of Organic Electronics in Callaghan, New South Wales, Australia. Scientists there have created photovoltaic solar cells which are very thin, made of natural organic polymers, and can conduct electricity. As they are liquid, they can be printed out on a roll and may be just the low-cost solution the planet needs right now.

Another truly exciting and potentially world changing invention are the solar panels and films made of vegetable waste that absorb stray UV light and convert it to clean renewable energy. Whole buildings in the future could be giant vertical solar farms. I've been eagerly following this amazing invention, called AuREUS, from Carvey Ehren R. Maigue who was the winner of the James Dyson Sustainability Award 2018. I can't wait to see its application all over the planet. According to Carvey himself, *'AuREUS could become part of our clothes, our cars, buildings and our houses.'*

We don't need plastic bottles!

Trials for a plastic bottle deposit return scheme in some UK supermarkets have been successful. As I've already mentioned in reference to a wider scheme, it won't stop single use plastic production but at least it will help the littering situation and put more money in our pockets. We can then support plastic free enterprises if we wish. Ideally, we need to cut out plastic bottles altogether. We should also be aiming for chemical free homes and businesses.

We've been indoctrinated by plastic and advertising corporations into using many substances and chemicals, often in pouches or spray-bottles. The average person uses approximately 30 plastic bottles of cleaning products a year. Gallons of chemicals go down our plugholes and continually enter waterways and oceans. These do no good to mankind, animal-kind, the insect population or the environment in general. If you or I poisoned someone we would soon be behind bars! Unscrupulous corporations twist findings, using what is termed 'mercenary science', to evade regulation to the detriment of a healthy existence on planet Earth. They rarely pay the price for the damage they do and often wriggle out of any law suit. We need to know exactly what is going on in these big corporations that should depend on good solid scientific evidence, and not spin.

When I was born, my dear parents were poor. There was no such thing as a disinfectant/surface cleaner except for Carbolic Soap wrapped in waxed-paper. My mother used to boil water in a kettle or saucepan on the gas stove and dampen an old cotton rag for cleaning around our large flat. In spite of this we all kept well with healthy immune systems. We urgently need to avoid harmful chemicals from now on and especially those in plastic.

Think planet first, not plastic!

Change is coming but we need to speed it up urgently

Pushing back against the greedy powers that don't want to change is essential. We may be forced to return to a more agrarian way of living on and with the land using regenerative agriculture and agroecology/agroforestry with sustainable land-use. The big corporations won't like it but we must roll up our sleeves and get on with the job, eliminating anything that damages the environment. We must also pester producers and retailers to look into alternatives for plastic products/packaging. One of my pet hates is the triggers on cleaning products/air-fresheners/antibacterial sprays that have to be binned because most are comprised of both plastic and metal. Many end up in landfill and billions are littered around the world. This is totally unacceptable!

Plastic carrier bags

These are one of biggest scourges on the planet. Bans and taxes are being implemented in many countries but only really work if there are good alternatives already in place. Fortunately, there are for carrier bags, but the best strategy is to avoid a bag altogether for small purchases. A change of habit is required by all. No-one needs a bag for one or two small items during a quick shop.

I've seen one toothbrush being put into a carrier bag at the checkout. Ridiculous, wasteful, and completely unnecessary even if the bag is paper! Most of us have pockets or handbags! The vast majority of plastic carrier bags are just there to advertise the seller anyway.

Plastic bag usage has significantly dropped in Turkey since a new regulation was introduced in January 2019. According to the authorities, approximately 150,000 tons of plastic waste was saved from landfill in an 11-month period. That's a lot when you think of lightweight plastic.

Thailand introduced a plastic bag ban in major stores in 2020 and Mallorca banned them in March 2021. Any ban will have a huge knock-on effect on the plastic industry. A year after the 2019 ban in New Zealand, 1 billion plastic bags were out of the system.

I can't wait to see the bans and taxes nibbling away at plastic year by year with more countries joining in. We don't need plastic carrier bags at all if we remember to take/bring our own. Some UK supermarkets are also no longer selling the notorious multipacks in plastic either because there was such a public outcry about them. People Power works. We should use it!

Ultimately, producers/retailers will be forced to find alternatives for the sake of the planet. The more we buy, the less they want to change. We need an end to petrochemical plastic. There are already alternatives they could use. All supermarkets need binding nationwide plastic reducing policies with more 'nude' produce and no 'Best before' dates. (The aforementioned Eyeball, Sniff and Quick Fingertip Taste Test used to work fine.) We also need a National Recycling Policy that all UK Councils have to adhere to, doing away with the vagaries of market forces driving decisions. Earth can't afford recycling centres closing because they have become less profitable.

Another problem to address immediately is 'Wish cycling' – recycling something that we think can be recycled but in fact can't be. A lot of plastic and plastic mixed materials come into this category. Plastic is so 'sneaky' that it can be made to look like anything now. Some 'paper' is actually a combination material, and can't even be separated for recycling. Cereal bar wrappers, soup and spaghetti meal pouches are prime examples. As is some 'cardboard' that has a plastic coating. Rattle packets that might have plastic wrappers inside and avoid them if you can. After use, tear any 'paper' pouches to see if there is a plastic inner layer. If there is, if possible, avoid them next time you shop.

Always check the back of products for the correct recycling information if they have it. Crush any 'tin-foil' packaging in your hand and if it springs back, it's plastic and can't be recycled in the normal way. There are specialist companies that can recycle this type of material but this is not the answer to the worldwide plastic crisis. Consigning single use plastic to history, is.

We should avoid throwing away any hard plastic we can still use. We must give it a good life as it has taken huge resources to make. When it 'dies', we can then responsibly dispose of it and replace it with something more natural and plastic free.

There is so much unnecessary plastic. Take plastic shoe-covers used in swimming-pools for example, that are just adding to landfills worldwide, and plastic windows in Spaghetti and Couscous packets. We certainly don't need a plastic window on any product when the packaging itself tells the story. Thank goodness some companies are ditching those windows now. It's such a simple change that will save money. There is no extra cutting or buying of plastic inserts, renders any cardboard packaging recyclable, and will make a world of difference to the environment.

At the beginning of 2020, a potentially world-changing lawsuit was filed in California against some of the top polluting companies as well as several other major food and beverage companies. A couple of well-known environmental groups are suing for damages to repair the mountainous problems of plastic waste caused by these corporations. They argue that plastic isn't just a global pollutant but an actual design fault.

There are organisations and charities designed to help reduce plastic by creating circular economies but this won't stop plastic being produced. Ancient Earth **will** survive whatever we do to Her. **We** might not if we don't do something proactive **now** to secure Mankind's future!

The next big question is, how do we live without plastic on what has rapidly become Planet Plastic?

Chapter four

Plastic free?

The desperate cry on every Eco Warrior's lips: 'How do I go plastic free when everything is plastic or wrapped in plastic?!'

Do a Bin/Waste Audit and a Pantry/Kitchen and Bathroom Audit.

Ask these questions:

- Do I/my family/business really need this?
- Could I use a reusable item instead?
- Can I cut it out of my shopping completely? (Something to consider if it's a nasty chemical product in a plastic bottle or spray bottle.)
- Is there a more natural sustainable product I could have bought in carboard?
- Is there a better packaging or no packaging option in another shop or supermarket?
- Could I buy this item in a reusable tin/jar/glass bottle?
- Is there a plant-based eco-pouch of one of my regular cleaners/antibacs available that I could use in an old clean spray-bottle?

- Is there a market or independent shop nearby that I could go to for bread and cakes, fresh fruit, vegetables, meat, etc.?
- Would I be able to make this without using a packet/pouch/sachet?
- Could I cook/bake this instead of buying it pre-packaged?
- Is there a zero waste/plastic free shop I could use for household/refillable products?
- Could I make/create/sew this instead or ask someone to do it for me?
- Could I buy something plastic free online to replace this if I can't find it locally?
- Is there a fruit and veg box service/online service I could use?
- Is there a milkman nearby who could deliver plastic free?

Some people think they have to monitor their plastic waste for a year. All well and good as an interesting project but why not cut it out in the first place. Its disposal wouldn't be easy anywhere in the world at this time. A month is quite long enough to monitor and change any plastic habit.

Just do what you can. We can't do everything but every little change helps us and the planet.

Some companies are taking the crisis seriously, or being forced to by their customers. There are supermarkets that use a non-plastic, plant-derived, water-based coating that creates an extra 'peel' around fruits and vegetables to reduce the rate it spoils. This potentially triples the shelf-life. Some are using laser-marking instead of sticky plastic labels. Being more creative about not using any plastic bag/film or label will certainly help save the planet. Each small change, be it domestic or commercial, soon adds up. We don't want to be recycling single use plastic. We need to completely replace it in everyday life.

Our precious soil

What I find worrying about recycling anything by way of our food collections here in the UK, and mostly incinerating waste which may have organic material in it, is that very few organic substances are going back into the soil to replenish it. The little that does end in landfill is then starved of the oxygen it needs to break down. As this is almost a planet-wide practice, we can clearly see that there is no benefit to Earth now. We can't keep doing this. Our surprisingly thin layer of soil needs constant feeding or it will die!

Astoundingly, every teaspoon of healthy soil on Earth has billions of organisms such as bacteria, algae, fungi, protozoa, nematodes and more that are essential to the health of Earth and us. We have evolved together. Sadly, many of us have lost this vital connection to the soil that keeps our immune systems healthy. We need to play in the dirt as children to boost our immunity, and become gardeners, even if we only have a small outside space.

If farmers rotate and diversify their crops and animals, don't poison the land and water in any way, including any plastic pollution, and also minimise any soil disturbance and add good organic compost, the soil does become more resilient and productive. It is also able to sequester more carbon and hold more water. Governments worldwide must support this way of farming and not greenlight more landfills and incinerators that do nothing but poison our precious Earth.

We must feed the soil, not kill it! Good healthy soil is essential to a thriving planet! Biodiversity is bio-security/ food security. Every tiny creature is **essential** to the health of Mother Earth. We must learn to tolerate them more. Please don't kill any that are not doing you or your pets/animals that you keep any harm. Remove them instead. There is a worldwide Insect Apocalypse happening right now because of intolerance and the global use of chemicals/poisons.

Healthy soil, healthy Planet!

A plastic free journey

I didn't know anything about the global plastic crisis at the time I started my own plastic free journey. I was just aware that there was a lot of plastic waste in my own life. There wasn't any media attention focused on it as it was before the BBC's *Blue Planet II* and before the Five Gyres were discovered. These are floating rubbish patches the size of countries in our oceans that are getting bigger by the day! The largest one, the Great Pacific Garbage Patch, halfway between Hawaii and California, now has a champion in the form of Boyan Slat and his Dutch non-profit company, The Ocean Cleanup. They're using the natural ocean-forces to catch plastics with a cork-line boom and a 2,000-foot-long floating tube that skims the surface.

According to the experts and those who have swum in the patch, like 52-year-old Frenchman Ben Lecomte, who set out on a boat from Hawaii to spend 80 days swimming through it, it's more like a 'plastic soup' of floating rubbish. The non-plastic detritus degrades and falls to the ocean floor but the plastic remains on the surface, unsightly, polluting and creating hazardous conditions for marine life.

A team from England's Newcastle University, led by Dr. Alan Jamieson, discovered a new shrimp-like crustacean in one of the deepest places on Earth, the Mariana Trench 7km below sea-level and underneath the Great Pacific Garbage Patch itself. They were so concerned at finding microscopic bits of plastic in the creature's stomach that they named it *Eurythenes Plasticus* to highlight the global plastic crisis. This research, with its disturbing findings, was supported by the World Wide Fund for Nature and has been published in the scientific journal *Zootaxa*.

'To truly rid the oceans of plastic, we need to both clean up the legacy and close the tap, preventing more plastic from reaching the oceans in the first place. Combining our ocean cleanup technology with the Interceptor™, the solutions now exist to address both sides of the equation.' - Boyan Slat, founder and CEO of The Ocean Cleanup

The Interceptor, a solar-powered barge, launched by Boyan Slat's company in 2019, is the first scalable solution to intercept river plastic pollution. It can be deployed all around the world in the most polluted river-systems, like the Klang River in Malaysia, where it has already been a great success. You can follow the development of these systems on https://theoceancleanup.com.

Other companies are using innovative designs to tackle the problem. John Kellett, the founder and president of Clearwater Mills LLC, installed the Waterwheel Powered Trash Interceptor in the Jones River in Baltimore, Maryland, USA earlier in 2014. 'Mr. Trash Wheel,' as the googly-eyed device has been named, uses booms to scoop the rubbish into its 'mouth' and a conveyor belt to lift it clear of the water and slowly deposit any debris into a skip/dumpster on the back. Water-wheels power the conveyor-belt with the addition of solar-powered pumps. 'He' has been so successful that they are employing 'him' in more rivers in the USA.

4ocean.com have also deployed their Ocean Skimmer in areas where there are accumulations of floating rubbish.

Fleets of ships to clean up floating detritus are also being considered by other companies. The idea is to incinerate it onboard with no emissions using pyrolytic gasification. This is where waste, including toxic/hazardous matter, is turned into a gas in an oxygen-starved atmosphere at very high temperatures. The clean safe energy created would then be used to fuel the electric engines.

Thankfully, many cities are already installing booms in their rivers and estuaries to capture plastic before it washes into the sea. This could also be incinerated in a land-based site, as they are starting to appear all over the world as one of the many solutions to the waste problem. Biochar is a safe by-product from the process that can be used as a non-toxic soil-fertiliser and to sequester carbon from the atmosphere. All aspects of this make me feel hopeful for the future as it could be the end of landfill.

It's heart-breaking to see beautiful countries destroyed by plastic waste. Recent studies revealed that the River Thames, that flows through the City of London and out into glorious countryside, is the most plastic-polluted river in the world. It has the highest recorded levels of microplastic particles, that endanger all life in this picturesque river course.

This is a shameful indictment on our way of life here in the UK!

We can't let Earth become a floating garbage-patch with land in the middle. It is heartening, though, to know that something major is happening. Now that drones are being used to track plastic it will be easier to see the route it takes from river to ocean. Any plastic removed as a consequence is a win for the environment.

No petrochemical plastic from production to sale to final disposal is ever good for the planet!

I started my Facebook page, and group, Yes, there are Alternatives to Plastic, to explore the possibilities of a plastic free planet and to share it with like-minded people intent on healing us and the wider world.

My page is about a lot more than just plastic but I want to focus on the scourge of Planet Plastic in this book because it worries so many when we feel there is nothing we can do.

The truth is, there is a lot we can all do.

It starts with us and our ingrained habits!

Chapter five

Land matters

If we want to save our oceans, we must first look at the land.

I'm encouraged when I see the good citizens of the world finding real practical alternatives to plastic. I'm sad that the large companies that have the wherewithal to do likewise are being slow in implementing the necessary changes we so desperately need.

'Our UK supermarkets have tremendous buying power and can influence brands to stop using toxic indestructible plastic.' - Sian Sutherland, co-founder, A Plastic Planet

One supermarket chain, even a large one, won't make much difference but all worldwide certainly will.

Modern plastic factories and their chemical partners do a great deal of harm to the environment as they contribute to climate change with their toxic emissions. Companies that make fizzy drinks burn fossil fuels to generate carbon dioxide for their products, adding further to the climate crisis. There are plenty of alternatives to soft drinks in plastic. We don't need to always reach for the plastic. There is glass that can be infinitely recycled.

I think plastic is the biggest manmade Frankenstein's monster on the planet at this time! Creating jobs that harm the planet is not acceptable and nor is poisoning the communities surrounding these vast sites that greedily swallow up land, watercourses and other natural resources. Plastic rules, and the plastic industry will prosecute anyone who stands in its way or speaks up about the dangers of living in the shadow of their factories.

'If it's in our marshes, it's in our oysters, it's in our fish and it's in our dolphins. And if there is plastic in them, there is plastic in us.' - Caroline Bradner, the Land, Water & Wildlife Project Manager for the South Carolina Coastal Conservation League

This kind of research that turns up such terrors makes me more determined than ever to get the plastic tap turned off. At least get the process started while I still draw breath. There is nothing good about single use plastic. This petrochemical product is destroying us and the planet. We just can't accept it as normal anymore. The answer to any new product, toy, whatever, can't be plastic and we must avoid buying it whenever we can. We must mobilise against an end to its dominance in our lives. Thankfully, more alternatives are turning up daily. We can also learn to listen to our grandparents and great grandparents and to wise members of Earth's indigenous cultures.

I'm so attuned to spotting plastic in the environment now. I once thought I saw a large piece in one of my plant pots in the back garden. I didn't expect it to be a mummified fish-head dropped from my flat roof by one of our local sea gulls. It really resembled a piece of 15-year-old greying plastic until I saw its teeth and eyes, and rapidly chucked it onto the garden from my trowel. At least this fish didn't have any plastic in it. It just looked like the nasty stuff naturally. (I can be forgiven that idea as fish-skin bioplastic is now a reality.) I saw something drop from the roof the day before but didn't take much notice. I now dread to think what is on my roof during the breeding season in my coastal town. I sincerely hope I don't see any plastic that has been brought in for food for the many youngsters that hang around up there. All-natural fish-heads, I can cope with.

No more plastic bags!

Plastic carrier bags are a blight on the planet! The loss of revenue to the plastics industry is the only thing preventing sweeping global bans. Any caring person can see these sorts of bans, along with taxes, are needed worldwide. We must remember to take our own reusable bags and containers when shopping, and buy in bulk when we can.

See the **Swaps – Thinking Planet first** section for ideas and materials.

There should be a sign on **all** shop doors:

'We love the planet and are not supplying any more bags. Bring your own.'

It wouldn't take long for it to spread. Customers would only forget their bags once after the difficulties of not bringing them.

Interesting solutions to plastic pollution:

- There has been great excitement about so-called 'Bioplastics' made of plants from both land and ocean.
- Plastic-eating mushrooms have been found.
- Argentine butterfly caterpillars, Wax-worms, that naturally eat the wax in bees' nests, eat the plastic that is Winter protection for hives with no harm to their reproductive cycle.
- Plastic-eating bacteria that can process difficult to recycle polyurethane have been discovered, but it could be many years before these can be used commercially.
- Mealworms that can digest plastics with no harm to themselves or the food chain have been found.
- There is also a process for clearing oil, called Bioremediation, where bacteria consume the spill, that could possibly be adapted to plastic pollution but again it is early days.
- Plastic is being reprocessed and turned back into fuel-oil.

There is far too much plastic for these to be viable options at this time. What concerns me is that if these do work, we will have Plastic Planet forever.

The plastics industry will just keep pushing it out and producers of goods won't bother to find alternatives.

Taking it to a ridiculous conclusion, we will just end up with more plastic, with mushrooms, small wriggly creatures, beetles, bacteria and enzymes trying desperately to munch giant landfills full of it! Why should creatures taken out of their own natural environment be expected to clean up our plastic mess?! The problem is there is too much of it and too few of them!

Airlines are at last addressing the huge problem of plastic waste and some have even handed out steel water-bottles. Cruise lines urgently need to put an end to dumping their waste in the oceans, which has added so much to the 5 oceanic garbage patches. There should be eye-watering fines if they are caught doing this, and if they become serial polluters their licence should also be revoked.

We're trying to save our one and only Home. Nothing should be more important!

I can't stop myself mentioning single use plastic pollution wherever I go, and recruiting as many eco warriors as I can. (I'm becoming quite the 'Plastic bore'! Don't get me started!)

Many people don't understand what's going on, and a lot don't care as long as they can keep their cosy lives of convenience. We can't stick our heads in the sand about this in the hope that it goes away. It won't! There is no 'away' when it comes to plastic! We breathe in plastic every day from the microparticles that constantly surround us from food, clothing, bedding, furniture, technology, gardens, and the things we touch daily. A new discovery is that plants can absorb plastic through their root-systems, and so we're eating more of them in our fruit and veg. I find this absolutely terrifying! God only knows what plastic chemicals/poisons have done to people of my generation and older who have had it in their systems for decades!

Remember, it has been calculated that each person on Earth ingests a credit-card-sized amount of plastic per week, mainly from drinking water, and a solid plastic coat-hanger-sized amount in a year. This is horrifying! No individual or organisation will ever be able to fully track microscopic plastic that endures on Earth over the years. We truly are Planet Plastic now!

Going plastic free is a minefield of problems for anyone who cares for Mother Earth but we must do it for the future health of all life.

We must also listen to Earth's indigenous people, who I believe hold the key to a safe and thriving natural world.

Chapter six

Learning from the past

I am proud to say that I am deeply spiritual and connect to ancient and wise Lakota energies and know that we are connected to and affected by the action or inaction of others. I feel so much part of everything that it hurts that we are so out of balance at this time. We must realign with Nature in order to heal ourselves as well as our glorious planet. Nothing I do is in isolation from my spiritual life. I write from my soul.

The native American peoples knew how to live with Grandmother Earth, never taking more than they needed to survive. Their lives were full of deep reverence for Nature's manifold gifts. They blessed the animal spirits they relied on in harsh times and thanked them for laying down their lives to feed and clothe them. To indigenous peoples, all things gifted by Nature have a use. Everything has a spirit and a purpose in being on Earth and is blessed and thanked to that purpose.

When a Lakota Warrior lost his horse, he honoured it by making a Spirit Horse Dance Stick in his horse's likeness. They loved and respected the trees, their 'Standing Brothers', and talked to and blessed the waters because they knew they were alive. The Lakota say, 'Mni wichoni', 'Water is life.'

Dr. Masuru Emoto in Japan carried out hundreds of experiments, before he passed in 1994, on the effect of words on water crystals. Positive and loving words made the crystals smooth and negative ones turned them sharp and jagged. He discovered that talking to water did have an effect on how it behaved. He was sickened to see what we had done to the planet's water-systems, how we had allowed them to become polluted and poisoned for financial gain. His books on these unique experiments are fascinating and a must read for anyone who cares about the health of water and, of course, we ourselves are composed of around 60% water, living on a water-planet.

The native peoples named their children after natural things and the animals that surrounded them, and their seasons by how they felt at that time of the year. They never abused Earth's treasure's but utilised them to bless and heal. They knew the now long-lost secrets of treading lightly on Earth. It is the tribal peoples of the world who are going to save it. We have moved too far from the natural way of living with Mother Earth and are paying dearly. All creatures have their own intelligence/survival instinct, that we need to learn from. We must return to a more natural way of living if we truly want to save our one and only planet from the ravages of greedy mankind.

I know we can't turn the clock back to these times but we can learn from them.

Now I want practical answers for how to live as part of Earth and not apart from Her. Single use plastic is a plague on our planet and would sadden ancient cultures like these who threw nothing away and used every single piece of material or flesh they could find a practical use for. Absolutely nothing went to waste.

I once had the sobering message from one of my Spiritual Guides:

'The fish can't breathe!'

Thousands of fish and sea-creatures are not only dying for lack of oxygen but becoming ensnared in plastic. He also warned us not to be *'a disease on the Mother'*. Sadly, I think this is what we have become.

I had the rare opportunity of working on an Ancient Iroquois archaeological dig in New York State in 1976, when I was an exchange student from my teacher training college to University of New York, Oneonta. Of course, what we found was completely natural: hammer stones, flint arrowheads and knives, and animal bones from hunting trips. We also found post holes for their long-houses. I found an arrowhead that had, according to the State Archaeologist leading the dig, probably been made by a small boy at his grandfather's knee. I felt deeply moved and privileged to have found such a wonderful piece of the past.

Ancient people's middens, rubbish heaps, biodegraded. Ours don't, and will be part of our vital thin layer of soil that blankets Earth and a blight on it for centuries. As I mentioned earlier, we must work for the soil and all its vital organisms and not against it. We have already done the unthinkable. We have abused beloved Mother and dishonoured Great Spirit who gives us life. We have violated our Home and continue to do so without a care for the consequences. We are selfish and no longer think of others or our magical planet. We have become a lazy, acquisitive, and dirty world full of things we don't need, and it is our arrogance that is killing us.

Of course, the planet **will** survive. We **won't** if we don't do something to restructure our modern lives of convenience and rampant consumerism. Earth nourishes every part of our lives and yet many are blind to Her charms. We can't let our ever-growing cities continue to destroy our world, the homes of high demand, financial excess and greed. We need a new way to live and we desperately need it, **now!**

'Only when the last tree has died and the last river has been poisoned and the last fish has been caught will we realise that we cannot eat money.' – Chief Si'Ahl of the Suquamish and Duwamish tribe, known by the Anglicised name of 'Chief Seattle'

There is nothing more Spiritual than being loving and caring Guardians of Mother Earth, our one and only Home while in corporeal form. We all come to Earth with a life purpose, whether we realise it or not. There are so many battles to join now that we have to pick one and go for it, and then when that is finally won, go for the next one.

I know that my battle is to help save the world from a tsunami of plastic that is engulfing the entire planet. I put myself here at this time to help heal the world from the damage that plastic has done in such a short space of time. I'm fighting for a more natural and plastic free planet that nurtures all life once more. I know that my spirituality helps me to connect and empathise more deeply with Nature.

Freeing ourselves from plastic pollution is one of the ecological bandwagons we should all be on if we want a clean and safe planet for our children and beyond. We can't keep wrapping things in plastic, specifically not food. We know that it photodegrades in sunlight in the wider environment and turns into microplastics, plastic pieces smaller than 5mm. It ends up as massive global pollution. To give you an idea of the actual size of pieces termed 'microplastics', the round studs on top of the iconic building-bricks are 4.8mm across. If pieces are larger, they are referred to as 'macroplastics'.

Two alarming materials have appeared on beach-cleans:

- **Pyroplastics** - created after plastic is illegally burned somewhere in the world and carried on ocean currents to be deposited on far flung beaches.
- **Plastiglomerates** - stones that contain mixtures of sedimentary grains, sand, and other natural debris, like shells and wood that is held together by hardened molten plastic.

We are going to see a lot more of these two materials as the world warms. Both look like natural stone but are highly toxic to life, often containing lead, and found on beaches worldwide. They're the result of melting plastic waste on campfires or from high temperatures on beaches. The only way they can be differentiated from stone is if they are seen floating.

This is the result of our plastic world over decades and will only get worse if it isn't stopped. Sir David Attenborough warned in *Blue Planet II* that the world's oceans are under the greatest threat in history.

How sad that our gorgeous planet's biggest champion should live to see this monstrous abuse! We were warned about this years ago and most people carried on doing the same old thing, accepting and not even questioning. Now we have to shout it from the rooftops before any major producer/retailer will do something to stop the sickening pollution of our waters! He also said his heart was breaking. My heart, and the hearts of all who truly care for Mother Earth, is breaking too! Workable natural sustainable alternatives to plastic must be found.

Many scientists throughout the world are now urgently working on solutions. We must take heart at a time when we feel so impotent that we can cure the 'disease' we've passed on to our beloved planet. We need laws to limit plastic in general but until those laws are implemented, we must not sit idly on our hands tut tutting about the state of the world. We have to do something ourselves in our own homes and businesses. Any grassroots movement starts with us, and the carrot and not the stick works best when it comes to persuasion.

People Power works! Why is it always the NGOs (Non-governmental organisations) that do all the work? World governments need to work together. All must implement legally binding, iron-clad policies that will protect our precious planet. We are **all** a vital part of Earth and nothing good should work in isolation from the rest!

I'm delighted when I hear someone has started their own personal plastic-free journey. I was a teacher, and once a caring teacher, always a caring teacher. As I said before, my Facebook page and group are about a lot more than just plastic. They're about saving the planet. I have to keep things interesting and well-balanced, though, as I would go nuts if I had to post constantly about the terrifying world of plastic. I need to know that things are improving globally in all respects just to keep myself on the right side of sanity.

I wish so-called experts would stop saying that anything replacing plastic is worse. Other materials we could use are not destroying the planet right now. Petrochemical plastic, is!

Like many, I don't have the wherewithal to go completely plastic free at present but I have drastically reduced plastic in my life and will keep going on the journey daily. It should be viewed as a journey and not an exhausting sprint to be achieved overnight.

Please don't immediately throw out all the plastic in your home or business and replace it with sustainable alternatives.

You will only make the problem worse by adding to the waste crisis. Use anything plastic, especially hard durable plastic, until it 'dies' and then replace it with natural materials if you can. There are so many now, that I explore in this handy book. You can dip into the **Swaps** section whenever required for years to come.

In Japan children are taught from the start to clean up after themselves at school and in the home, and it's to do with respecting each other and the environment. We desperately need to follow this ethos in the West. What experts often fail to mention is that the human race is infinitely adaptable given the opportunity.

'We are starting to now realise that the notion of everything being disposable... is no longer viable.' - Andrew Morlet, Ellen MacArthur Foundation

The European Commission's new proposal is entitled the Circular Economics Initiative and will include extending the life of products through reuse and repair. It will also champion affordable upgrades and extended software support systems for any tech.

A Green Circular Economy is the way forward.

So, let's adapt now!

Chapter seven

General advice

Alternatives to Plastic Mantra

Don't be discouraged by the 1 piece of plastic you put in the bin today. Celebrate the 99 others you didn't!

Well done for everything you're doing already!

Chin up!

We're winning!

The situation worldwide is truly concerning at present. We mustn't let it overwhelm us to the point it affects our health. Just do what you can when you can.

I especially want to thank all the litter-pickers and beach-cleaners who are doing an amazing job as well as raising awareness of the horrendous global problem of beach plastic.

'Every piece of plastic we remove from the beach is a victory for the ocean. Every piece we stop from being produced is an even bigger victory for the planet.' - Hugo Tagholm, CEO of Surfers Against Sewage

We also need to push the water companies hard to stop using microbeads to clean sewage, that often end up on beaches. Plastic should not be the answer to anything now! Why ban microbeads in cosmetics and toothpaste in the UK in 2018 and not just ban microbeads outright for the danger they are to all aquatic life and us?

The path for the plastic-free warrior and zero-waster is difficult so we must be gentle with each other. We want to see a cleaner and safer world for all living beings. We have to be realistic and accept that as long as there is oil there will always be some form of petrochemical plastic. What we have to concentrate on, and eliminate as soon as possible, is ephemeral single use throwaway plastic that is suffocating and poisoning the planet.

Be in no doubt that plastic waste is highly toxic to bodily functions and a constant hazard, trapping and killing animal, sea and bird life.

Scientists working in the Pacific and Indian Oceans have also found that it's changing the chemistry of some seabirds' blood. Microplastics are also found in the gills of fish and other water-life. They are breathing them in as we are.

Don't forget that plastic in water is the perfect breeding-ground for bacteria that latch onto it. According to research carried out in Northern Ireland, almost all of the marine plastics studied harboured antibiotic-resistant bacteria which may be spreading, and is therefore even more concerning.

When gathering waste, it is advisable to go out with a well organised group that has all the right equipment but if we do it alone, in a family group or with friends:

- It is essential to wear protective gloves, especially on beach cleans. On some cleans 2 pairs together is advised.
- Take several strong waste bags.
- Avoid picking anything up with bare hands and use a strong litter-picker where necessary.
- Have a 'Sharps-box' handy to put sharp objects in to avoid injury.
- Avoid the use of synthetic hand sanitiser if possible. According to scientific research, it's hormone-damaging gel in plastic, and often contains high levels of microplastic that end up in waterways and oceans. Organic and Vegan sanitisers are available from eco businesses. Beautykitchen.co.uk does a refillable one in an aluminium bottle that is decanted into a small hand-spray, and is far more pleasant to use on a daily basis.
- Avoid Wipes that are 84% plastic and are no good for us or the planet.
- We must remember to wash hands thoroughly with hot water and soap as soon as we can or when we get home if we can't sanitise them effectively while out and about.

Do we really need chemicals in plastic?

Under the kitchen sink
In the bathroom
In the garden
In the garage
At work

We must use up whatever we have first, as disposal is not easy and keep in mind that **chemicals should never go down the drain**. Replace with non-toxic natural alternatives. You may even find that you can cut out a chemical altogether with one of the handy tips in the **Swaps** section of this book.

Everyday shopping

This is the biggest headache at present but there is a simple way to start thinking plastic free. Once this becomes a habit, we will never again think plastic first. We will happily explore the many options available and constantly look for natural and sustainable substitutes for all plastic products.

Step one

Remember, at home, do a Bin/Waste audit and a Pantry/ Kitchen cupboard audit. At work, do a Waste bin audit.

- Examine the plastic and plastic packaging and film for each item you're throwing away. Make a note of what you throw away and why.
- Don't overwhelm yourself with the seemingly monumental task. Just pick one item per shopping trip to swap.
- Avoid anything that has a plastic-wrap label as well, such as some sprays and bottles of drink.

Remember: Use any hard plastic until the end of its life, dispose of it responsibly, and then replace with natural sustainable alternatives.

Step two

- Shop with reusable bags.

When it comes to not using a plastic bag for convenience, sadly, most men are difficult to target. So, it is up to we women to encourage them to take a reusable bag when out and about. This is one situation where nagging is a necessity. (Sorry, gentlemen!)

Keep bags by the door and in the car so you are always prepared. You can even leave a sign near the door:

'Got your bags?'

There are many natural carriers available:

- Cotton tote/produce bag – will last for decades and can be thrown in the washing-machine with the weekly wash. Look for the Global Organic Textile Standard (GOTS) for everything made of cotton to ensure sustainability and best practice.
- Jute bag – some bags have a plastic lining to make them shower-proof, undoing all the good of the natural Jute. 100% Jute bags are made of thicker material and are more expensive on the whole. From past experience with this type of Jute and thin plastic material, the lining only stays in one piece for about 4 years' use, then starts to shred. Go for unlined bags if possible.
- Cotton string bag – strong and what we all used decades before the plastic bag takeover.
- Willow basket – strong and wonderful for shopping for fruit and veg.
- Straw bag with an organic cotton lining.

- Old coffee sacks - turned into reusable bags. Perfect upcycling and no plastic lining.
- Use boxes from the supermarkets to carry things to the car to avoid asking for bags if you forget yours – also what we did years ago.
- Pillow cases make good large bread bags if you're shopping at a traditional bakery that doesn't wrap in plastic.

One way to avoid buying anything special for the job is to make your own tote/produce bag out of an old tank tee, so you have readymade handles. Cotton tees are best. Charity/thrift shops are a good cheap source where you also find men's tees for the gents in your life who may not appreciate a pink or flowery bag, although there are a lot of enlightened men now who wouldn't mind at all.

Sewing machine method:

- Turn tee inside out and sew the bottom up, or get a family member or friend to do it for you if you don't have a machine.
- Turn right side out and you have a perfect bag.

No sew method:

- Anybody can do this.
- Flatten out the t-shirt.
- Cut a fringe on the bottom.
- Stretch the fringing and then tie opposite sides.
- Perfect bag.

If a shopping trip is a major weekly production for the family, then there are concertina-type shopping bags that go into the trolley that can be easily loaded into a vehicle, avoiding all the extra bags. Or use boxes, available from stores, and keep those ready for action. If you already have any large plastic containers, use those in the car, for your plastic-free shopping, if you have the capacity.

Worryingly, basic hygiene went right out of the window when plastic film, gloves for handling food, and hand-sanitisers came in. One habit we **must** return to is washing our hands more often, with good old bar-soap and hot water. Regular hand-washing used to be common sense but has been forgotten because most things have been wrapped to an inch of their life. We will be going back to handling a lot of 'nude food' once stores have plastic free produce, and therefore must use clean hands.

If our hands are dirty, domestic hand-sanitisers will not be effective as some labelled 'antibacterial' contain only antibiotics, which will not protect against viruses and have been proved in some cases to be hormone-damaging. Alcohol-based ones are better but the alcohol content of most supermarket/over the counter ones is not strong enough to kill some bacteria. Natural ones are available that are just as effective.

The World Health Organisation states that washing hands thoroughly for at least 20 seconds with soap and hot water will dislodge bacteria and viruses. It also recommends that we wash our hands regularly, at least every 2 hours and before and after we go out. We don't want to transfer germs to others from items like metal objects and handrails that viruses can live on for hours.

We must make sure we wash hands thoroughly after going to the loo and after blowing our nose, coughing or sneezing, especially when out and about. Carrying our own bar soap in a tin or soap-box, and even a small towel, is good practice if we're worried about cross-contamination in public places. Back at home, we should wash hands again so nothing harmful is transferred to us or our loved ones.

This is all common-sense basic hygiene.

Single-use plastic, and other disposables, according to a number of reputable independent scientific studies, harbour viruses and bacteria, whatever pathogens they have picked up from production, transport, stocking in warehouses and stores, and final use by customers. It's pretty shocking when we stop to consider it. There should be more research carried out and solutions found as soon as possible. I think we have been far too cavalier about any potential dangers.

We should always remember to wash any 'nude' fruit and veg before consumption if it isn't going to be peeled because it might still have pesticide residue on it. One of the most proven effective methods to remove any is to soak fruit and veg that can be soaked in a teaspoon of baking soda, swishing it around for a few minutes, then rinsing with cold running water and patting dry on a clean tea-towel.

Washing hands thoroughly with hot water and soap after handling raw meat or fish is **vital** for safety. We must also make sure that chopping boards, knives and surfaces and **any articles** used for food preparation are kept spotlessly clean.

This is common-sense basic kitchen hygiene.

The best way to eat healthily, safely and plastic free is to grow as much of our own fruit and veg as we can in our gardens or on an allotment. You're lucky too if you have a friend with green fingers, and can swap produce with them. The easiest things to grow, even in a small garden, are fruit trees, on a dwarf root-stock so they don't get too big, and fruit bushes. Both can be bought inexpensively in garden centres or some supermarkets.

There is nothing like fruit from your own trees and bushes, chemical free, picked fresh, warmed by the Sun and plastic free. You will also save a lot of money. There are a lot of quick-growing vegetables that you can sow too, even in a window-box. There is a wealth of growing tips online and in gardening forums on social media.

Clothing

The clothing industry worldwide is a massive polluter and we should try to gradually transition our wardrobes over time to natural materials. Many of our clothes contain plastic. Only 100% natural materials are harmless to the planet on disposal. Any article that includes plastic will leave that plastic behind after the organic element has rotted. There was an horrendous photo doing the rounds on Facebook showing the plastic fibres left in a pair of stretch-jeans after the cotton fibres had been removed. It was still obvious they had been a pair of jeans.

Therefore, we must try to avoid:

- Polyester
- PVC
- Spandex
- Nylon
- Acrylic
- Elastane – found in underwear and most stretch-clothing, including so called 'natural' socks and jeans.
- Polyamide
- Metallised fibre – this can shed too!
- Mixed materials, even if they include something natural.
- 'Poly-something' - usually plastic.

Plastic microfibres hang onto odour-causing bacteria in a way that natural materials don't. Therefore, washing natural materials as frequently as manmade fibres is often not required. Each time we wear and wash manmade materials they shed millions of microfibres. These threads of plastic are so small that they drain out of our washing machines and pass straight through wastewater treatment plants and out into our rivers, seas and oceans. Research carried out at Newcastle University, UK, has shown that the 'Delicates' cycle, that uses more water than all the other cycles, seems to be the most important factor in dislodging microfibres from clothing. Therefore, when washing any 'plastic clothing' we should not use this particular cycle. When we wear these materials, especially the 'fluffy' types, we are in danger of breathing in the microfibres directly as well.

Once in the environment, plastic microfibres absorb toxic chemicals and bacteria. Sea creatures eating these fibres can pass them up the food-chain, called 'bioaccumulation', and into us via seafood like fish and mussels. We urgently need to find a more efficient way to make clothing without using any plastic or depleting natural resources.

No-one wants to be eating their clothes!

There are two new inventions to reduce the amount of microfibres entering river-systems and oceans: a fine mesh bag that catches microfibres that is put into the machine, and a Silicone ball that has been based on the way a coral catches food with its arms. Some washing machine manufacturers are starting to build filters into their machines. Other companies supply an add-on replaceable filter that attaches to any machine but in my humble opinion is expensive and wasteful and, ironically, is also plastic.

I think our clothes would last much longer without constant washing. When I was a child, and mums worked mostly at home, aprons were the choice for keeping clothes clean of marks and spills. You never saw my dear Mum or Auntie without their waist 'Pinnie', short for Pinafore, on in the house. I even reluctantly wore one on messy kitchen days. Sometimes my mum forgot to take hers off and I had to whip it off her before she left the house. It wouldn't have looked good if she'd arrived in the library office with one of her lovely homemade 'Pinnies' on. There again, she might have had some orders for them. Maybe we should make cotton waist-aprons, full bib ones for ladies and gentlemen, made from repurposed materials, normal again. They just go in the weekly wash and save our better and more expensive clothes from splashes and general wear and tear.

For all clothing, I recommend the 'Eyeball and Sniff Test' before washing.

When buying any cotton clothing, look for the Global Organic Textile Standard (GOTS) to ensure sustainability and best practice.

Avoid what we can, and plant the 'seeds' of change

90% of products on Earth are now plastic. We can't let Big Oil/plastic corporations run the world anymore. So, we the people must create a grassroots uprising. If we don't, they will win and nothing will change as they continue to violate the planet for profit.

Put good women, and I stress 'good', in positions of power and the woes of the world will improve overnight. Strong caring women are becoming world leaders with policies to save us and Earth. A lot of the burgeoning eco-businesses are run entirely by women and there are many co-operatives run by inspiring women worldwide. There are some wonderful men who are doing all they can to fight the monsters but, sadly, not enough. We must recruit more!

We need oceans free from pollution. Filter-feeders should be filtering sand, not microplastics. Anything dumped on the land does end up in a waterway or ocean. The wind blows it around and the rain washes it into drains and out into waters that should be thriving with abundant life, not struggling to breathe because of mindless pollution.

Once we cut out unnecessary chemicals in plastic, we have more money to buy things that matter. I've cut out so much plastic that I can afford some totally plastic-free items. I also buy the amount of fruit and veg I actually need. I avoid any packed in plastic whenever I can on my tight budget. If I can do it, anyone can.

We have to be soldiers fighting on the beach in the war against single use plastic to stem what has been called 'The Global Plastic Tide'. I do believe we will have the world we need. One that is free of manmade pollution. There is nothing like a big healthy expanse of water with an abundance of life to lift the spirits. We live on a water-planet and must look after its waters. What happens on the land **will** end up in the water.

We can all do something practical today. So, what next?

Chapter eight

Consigning petrochemical plastic to history

Remember: Not only is plastic in our waterways and oceans, it's in the very air we breathe, in our food, and now in our bodies!

There are bans on Polyfluoroalkyl substances (PFAS, now termed 'forever chemicals') used in food packaging because they have caused so many health issues throughout the world. PFAS chemicals are often used as water and grease repellents and are found in paper and cardboard that comes into contact with our food. The compounds can migrate into the food and us. They have also been used in non-stick coatings and waterproof fabrics for decades. Hardly any 'modern' substance seems to be safe for life because it often contains toxic chemicals and/or plastic. Mother Earth can't take any more plastic/toxic chemicals and nor can we.

'We don't own the Earth. We belong to it. And we must share it with our wildlife.' – Steve Irwin

Why I hate petrochemical plastic

One of the many reasons I hate plastic with such a passion is that it breaks my heart to see what it does to our spectacular wildlife. It kills not only wildlife but farm animals and even domestic pets when accidentally ingested. Aquatic life and seabirds are endangered when plastic 'Ghost nets', deliberately abandoned or lost, entangle and strangle. They cut flesh like a scalpel and drive whole species like the wonderful Vaquita (the smallest Porpoise) to extinction. Not only are they a terrible danger to wildlife but a trial to get rid of for the rescuers who often have to drag the heavy nets off the beach once landed. Then they need safe disposal. Some nets found on UK beaches are being turned into new products but to me that is just delaying pollution.

I always carry a small pocket-knife just in case I come across trapped birds or animals, especially as I live by the coast.

The global fishing industry must rethink nets and return to thicker natural ones to save our precious ocean life.

'Until he extends the circle of compassion to all living things, man will not himself find peace.' – Dr. Albert Schweitzer

We constantly underestimate the intelligence of animals and birds. We chose to do this in the past so we could cruelly exploit them. We're finding with more ongoing research that they all possess intelligence. They need to be respected as the sentient beings they are, important to the health and well-being of the planet.

Ocean-going birds are forced to feed their young with plastic detritus because their parents can no longer find enough fish. In Antarctica, Boluses, the undigested bits of food that Penguins and other birds regurgitate as part of the feeding process, have been found to contain plastic. Eggs may also be absorbing toxins from plastic brought into the nest. Studies in Australia have found that Hermit crabs that should be finding shells in which to live die if they choose anything plastic as a home. They often become trapped when they can't grip slippery surfaces to climb out. Turtles, those glorious ancient ocean-wanderers, think that thin plastic bags are jellyfish, their staple diet. Algae also grows on ocean plastics, further confusing and endangering any creature that ingests them.

The good news, for the ocean and turtles, from Indonesia is that the Cassava plant, which is abundant, is being used to make fully biodegradable/edible, even drinkable plastic bags that do no harm to people, wildlife or the environment. Being more expensive than the much-used plastic bags, it will be difficult to change the culture but I'm sure it can be done with the right investment and education, 2 more vital keys to a new world.

The study of Manta rays in The Galapagos Islands has revealed that they filter microplastics along with their usual diet of plankton. Even the microscopic plankton itself is full of plastic particles which the rays may be absorbing. Disturbingly, a team of scientists from Germany's University of Bayreuth have discovered that microplastics in water can grow a bio-coating that may enable them to infiltrate living cells. More research is being carried out to determine exactly how the cells are affected and what damage could be done.

Everything happening/changing in The Galapagos is now happening in the wider world. This is why scientists from around the globe are studying the unique endangered wildlife and its equally threatened environment.

It's heart-breaking to see images of majestic whales beached, dead, with stomachs full of plastic. We must appreciate the interconnectedness of animals and the health of the natural world. Whales are vital to a healthy ocean and facilitate carbon absorption in three ways:

- When diving, they push nutrients up from the ocean-floor to the surface where they feed the phytoplankton that generate half the planet's oxygen, marine flora that suck up carbon, as well as fish and other marine creatures.
- Their 'poo' enriches the ocean and introduces nutrients that allow marine-plants to grow in the area. As these photosynthesise, this process also furthers carbon capture.
- When a whale dies naturally and sinks, any carbon stored in its body goes down deep into the ocean where a huge variety of flesh-eating creatures safely dispose of its body.

Why do birds, domestic and wild animals and marine life feed on plastic?

Sadly, there is a simple answer. It smells and tastes like food to them. Some creatures also have poor eyesight so can't see what they're ingesting. Remember, plastic harbours bacteria in the ocean and is a fertile breeding ground for some deadly ones. We humans can do something to improve our own situation but other Earth-creatures can't without **our urgent intervention.**

'If your bathtub was overflowing, you wouldn't reach for a mop to clean it up; you would turn it off at the source. And that's what we need to do on plastics.' - David Pinsky, an anti-plastics campaigner at Greenpeace

Unless the plastic tap is turned off for good, our natural world and its wonders is doomed to a slow and agonising death!

It shouldn't be a frightening future for our children!

Our young people are rightly concerned by what is happening to our planet. We adults, and especially the scientists who are fighting for the preservation of all life, suffer from what is now termed 'eco-anxiety/eco-guilt' about the enormous changes that have happened to our beloved planet in such a short space of time. I have seen petrifying changes in my lifetime already.

Most of us only need to look out of our front window to see the result of our society of convenience and carelessness. We adults must set an example to future generations. Explain that you love Mother Earth and when others see you doing something positive and practical to help, they will follow.

When it comes to friends and relatives and present-giving, the main thing to do is stand your ground and don't give in. Once they know you don't tolerate anything that will harm the planet, they will respect you and your choices.

People treat us the way we let them. So, don't let them steam-roller your eco-choices for a better and brighter future.

I'm sad that I have to mention this but parents should teach their children to clean up after themselves. Mummy and Daddy should not be expected to do it because it's less bother and they don't need to shout at the children or become boring about asking them for the millionth time to tidy their rooms. If they keep doing everything for their children, they will perpetuate this terrible lazy global culture of 'It's not my job!' Oh yes, it is! We have a duty to the planet to keep it clean and safe for all life, and it starts at home, teaching good habits that harm nothing and no-one. Making sure that our own surroundings are tidy and don't look like the local landfill will instil a sense of pride in our families which will spread to others.

Act locally but think globally!

'Earth is not a platform for human life. It is a living being. We're not on it but part of it. Its health is our health.' – Thomas More

If we throw in the towel when it comes to plastic pollution our world will be nothing but Planet Plastic forever. The more I learn about plastic and how it is used in the food industry the more horrified I am, and the more determined I become to rid the world of single use plastic for good.

Food-packaging was the most common beach rubbish in 2018 according to beach cleans and product audits that took place in 120 countries. Now we have Smart Plastic that has electronics embedded into it. The potential for a huge boom in plastic is petrifying! Products might be able to tell us when they are low and need reordering. Convenience, greed and sheer laziness is destroying the planet every second!

I don't like circular economy when it comes to plastic

I just want to see an end to it, but for anything else it is the way of the future. There is a new idea being tested in the USA, the Loop Alliance, where companies send out their products in reusable containers that can be returned, washed, and reused. Somewhat like the milk rounds of the past. It keeps everything in a closed loop recycling system and avoids landfill or incineration altogether. Apparently, it has a lower carbon footprint than making things from scratch even though transport is being used. By April 2020 four hundred brands had signed up and are now developing new packaging for their various products.

One of the best ways to avoid plastic-packed fruit and veg in the UK is to participate in a glean in our fields and orchards, gathering good produce that has been rejected by the big supermarkets or is not worth a second pick by the farmers. This is then distributed to any organisation that can make use of it, such as homeless shelters, food banks, women's shelters and community food kitchens.

Feedbackglobal.org is one such organisation, that you can sign up with as a volunteer on Facebook or on their website. After a happy day gleaning in the fields or orchards you can take home some wonderful plastic free food straight from the fields and trees, and it's free! The farmers who sign up for the network are only too pleased that their fresh produce finds new homes and doesn't go to waste. Win all round!

Community gardens are also becoming very popular and the UK take-up of allotments is huge now! More and more people want to grow their own fruit, veg, and flowers without worrying about pesticides and plastic.

Paper versus plastic

Sorry, but paper bags are the lesser of the 2 evils. They are not suffocating and poisoning the planet by the trillions! Nor are there paper bag bans in many countries. Remember, in some parts of the world, there is a plastic sheet/bag layer in the earth.

We need to take our own reusables with us everywhere. Once supermarkets and shops realise we have changed our habits, they will stop supplying paper bags as alternatives to plastic ones.

As long as we can get it right, with sustainable managed tree and Hemp plantations, with owners who really care, and recycled paper, even old clothing can be turned into rag-paper, being used throughout the world, it should be manageable. There would be no damage to the natural world if we keep it all in balance and don't start more monocultures that crowd out much-needed diversity. Of course, there is always room for other natural alternatives.

We have to look at the way we recycle and what we recycle next if we want to save our poisoned planet.

Chapter nine

What should we recycle?

As you are now aware, I am totally against recycling plastic. I'm also against mining landfills for any, which has been suggested for the future. It's best for us as individuals with buying-power to avoid plastic altogether whenever we can. Recycling or 'selective remanufacture' of post-consumer plastic is never going to solve the global plastic crisis. Nor is making furniture, houses, roads, pavements and parking spaces out of recycled plastic. Ecobricking, or making clothes/rainwear out of recycled polyester is no better. Manufacturers will never bother to find natural sustainable alternatives because, as we all know now, 'Plastic can be recycled'. The only way to end the plastic crisis is to get rid of what is already here and stop churning it out!

In the case of plastic roads, plastic in the surface breaks down into microplastic that finds its way into soil and bodies of water when it rains or floods. These particles absorb other pollutants in the form of oil and rubber/plastic dust from tyres, that will continue to pollute throughout the life of the road. Just heating the plastic to start the process releases harmful toxins. These roads are long plastic timebombs from start to finish!

Ecobricks are timebombs in outer casings of plastic bottles. They may work in countries that have no choice but to use their mountains of plastic waste this way but it is not a one size fits all situation. Even then there are the dangers of the bottles photodegrading and breaking down just to drop all the plastic and building materials, like solid cement, back into the environment.

Where we do have a choice, we should not be doing it. Even scientists don't know the results of jamming together different plastics. Plastic pieces are so diligently collected and stuffed into the bricks for all the right reasons but sadly will degrade in time and become problematic wherever the bricks are used. Another reason eco-bricking doesn't really work is because 'Brickers' often don't reduce their plastic habit as 'It can just go in the brick'. On the plus side, however, some do reduce their plastic consumption after they see the enormous amount they've collected to put in their not so eco-friendly bricks. It's better to cut down on as much plastic as we can, eventually weaning ourselves off it altogether.

So many products are still in plastic. Nothing is eco-friendly if it is in or contains plastic.

I'm so tired of seeing recycling as a solution to the global plastic crisis. The only solution is to stop wrapping and bottling in it in the first place. Natural and sustainable alternatives must be found to **all** throwaway plastic. Recycling our waste is treating the symptoms but not curing the disease of mindless over-consumption.

Another excuse for recycling plastic is that no more virgin plastic will be produced. That is just not the case.

Big Oil in the USA is commissioning more plastic factories to 'meet demand' for the future. A demand we must stop in its tracks. They know that in future there will be no profit in fossil fuels as the world wakes up to the climate crisis. They will therefore need plastics to fill the gap in their revenue-stream. To me, their greed and arrogance is completely unfathomable at this tipping point for the planet!

Can we live in a mostly recycled world and what should we recycle?

Any petrochemical plastics – No.
They can only be recycled a couple of times until they are so degraded that they're no longer usable, ending in landfill or incineration for fuel. Any product made of recycled plastic/ocean plastic is just prolonging the agony of a plastic planet. If we end fossil fuels that are causing climate change, we get rid of plastic at the same time as they are needed in the production of all petrochemical plastics.

The small plastic pellets, called Nurdles, that begin all plastic products, are one of the world's worst polluters. Think of Nurdles as 'baby plastic'. They're washing up daily on our beaches, especially where there are plastic factories or from container spills in our larger river-systems and oceans. Bacteria sticks to plastic and that is why they are used in wastewater treatment plants, but this is what makes them such a danger in a watery environment if they escape. As these pellets also resemble fish-eggs in water, they're mistakenly ingested by many marine creatures. Worryingly, studies have found fish in plastic-polluted areas covered in lesions and cancerous growths.

'Plastic is fossil fuel in another form. Everything that happens before you see that plastic on the shelf is emissions intense. It releases all manner of pollutants and toxic chemicals. At the top level, dealing with the climate crisis requires dealing with the plastics crisis.' - Steven Feit, a lawyer at CIEL, The Center for International Environmental Law, and contributor to a report on the oil and gas boom

Tins, and cans in Aluminium – Yes.
75% of the aluminium ever made is still in use today, showing how easy and effective it is to recycle this valuable metal.

Metals in general – Yes.
Many different types of metal can be recycled.
We don't want to keep digging them up.
Check advice on what can be collected, and where, for recycling.

Paper and cardboard – Yes.
Strip off any plastic tape and bin it before recycling cardboard boxes. (I stick the tape to itself so that if it does get into the environment, it won't trap small animals and insects.)

At work/home office
Set your printer to two-sided printing to avoid waste or use the back of anything printed for writing notes, lists

and reminders. Recycle all unwanted paper and cardboard.

This is advice from my UK Council about recycling cardboard food packaging:

- It doesn't matter if paper gets wet as it's part of the recycling process when it gets to the plant anyway.
- Frozen food boxes usually contain contamination like grease/fat or food and should be treated as refuse and binned. For example, frozen fish boxes, pizza boxes, cake boxes, etc.

  You can tear off the greasy or contaminated piece of the box and put the clean piece in the paper recycling.
- The factories that process it do not want contaminated card.

Many fine new products can be made out of reprocessed paper and cardboard, even new paper.

Glass bottles/jars and whole glass – Yes.
Wash out and remove **paper labels** and recycle them if you can. It doesn't actually matter here in the UK if we can't get labels off jars to recycle the paper as they are burnt off in the recycling process.

I remember years ago you could buy a very clever inexpensive bottle-cutter and make your own tumblers from old bottles. Still worth a try if you can get hold of one now as some bottle-glass is really attractive.

Glass is still litter and a danger when broken! It has to go into the normal waste stream. Wrap it up in newspaper or put it in a cardboard box for safety. Especially do this with **broken mirrors** and **ordinary glass** as it is usually a different type of glass from bottle-glass and therefore can't be mixed with it for recycling purposes.

Electricals – Yes.
Components, including those made of gold, can be retrieved and reused. If an appliance has a plug, battery or cable it can be recycled in the UK. Check your own country's advice on recycling electricals.

Clothing – Yes.
There are many ways to reuse good material.

I recommend removing any buttons first and keeping an old-fashioned 'Button jar'.

Charity/thrift shops will take worn/torn clothing for ragging. Just kindly mark the bags 'Rags' and the shop staff won't waste any time sorting them out.

Food – Yes.
All good unsold food should be donated.

At home we can regrow a lot of fruit and vegetables from the pieces/roots we cut off and the seeds. There is good advice online on how to do this effectively. Many more people are now doing this to avoid food waste, as well as wasting money. Why buy when you can grow!

See **Food waste** for more.

Fruit and veg waste – Yes.
If it can't be eaten it can be processed to be put back into the soil-layer. Many fruits and vegetables can be made into substitutes for plastic and are now being explored by the food and beverage industries.

Anything other than plastic that can be effectively and safely reused, recycled/upcycled or repurposed should be.

Products that are made of natural materials that tell us to 'Take & Toss' or 'Wash and Toss' must be banned or we will never save the planet from our throwaway mentality! Even if something is made of Earth-friendly materials we should use it for as long as we can before disposing of it or recycling it. We must put an end to our 'disposable' planet. There is literally no 'away' for plastic on Earth except for incineration, which is far from ideal with its toxic emissions, or turning it back into oil.

Producers must make recycling easy for us. This is the way of the future. However, we can't be expected to spend a good 15 minutes per item scrapping off labels, that seem to be stuck on with superglue, and separating materials like plastic and metal hybrids. More thought needs to go into their creation. If something isn't easy to recycle, very few people will bother recycling it!

The Japanese village of Kamikatsu started its zero-waste journey in 2017, recycling all waste using a staggering 40 different categories. It was extremely time-consuming for all and they were hoping to be completely zero waste by 2020 but didn't totally succeed. What did succeed, though, was the creation of a caring society that helped to do what was necessary in a village with an aging population.

Personally, I think that's a win. Central governments and communities must work alongside each other if they hope to achieve their goal of a zero-waste world. All things that can't be recycled should be repairable in some way and it is the producers' responsibility to make this happen soon. It must be cheaper in the future to repair something rather than replace it. Sealed plugs on appliances should be outlawed so we can go back to repairing those as well.

Korea has started to recycle and compost its way to sustainability. They've introduced 'intelligent' automated recycling bins, that prevent recycling of the wrong item, and regular food waste collections as well as growing rooftop farms. We should be doing more here in the West to address our sickening and escalating waste problem.

Many people can't even bear to go shopping because the amount of plastic is so overwhelming. I call it the 'Plastic Planetary Plague Syndrome'. The straw that broke the camel's back one morning for me was a woman buying every chemical in plastic she thought she needed under the sink. My heart sank so low that it was mixed up with my socks because all other items in her trolley were not only toxic to the environment but in plastic as well. I was seriously in danger of exploding that morning and if one more person says, 'There is nothing wrong with plastic', I will!

If there truly was 'nothing wrong with plastic' we wouldn't be in this global nightmare right now!

I'm sick of Greenwashing, a marketing strategy that suggests that a product or firm is environmentally friendly when it clearly isn't. I can't stand all this 'recycled ocean-plastic' crap. It's still plastic and it will probably end in landfill. It will also have taken huge resources to clean off all the bacteria on it. Greenwashing par excellence!

There is a lot of Greenwashing going on now because the world **is** changing and everyone wants a piece of the pie. I swear, though, there is more plastic than ever. The plastics industries will win if we don't do something. I'm sure you too have noticed more plastic sachets, pouches, toys, kitchen utensils, sauce bottles, plastic film and wrapping on tins and bottles, sweets, chocolates, biscuits, snacks and nuts entombed in plastic. Added to that there are more plastic clothes and shoes, garden accessories like solar lights and planters. The list is sickeningly endless!

The fat balls I buy for the wild birds in my garden were in plastic, with plastic netting, for years. After writing to the company several times, they appear to be net-free at last. Maybe your email or letter could be the last straw for a company before they are forced to act. It is well worth pestering for change.

We don't have to buy what they're selling.

All the promises to change from plastic bags to paper don't matter a jot when your business relies on customers buying your plastic products or those wrapped in plastic. Shops supply them and customers will blindly buy them because societies globally have been indoctrinated into buying this way for decades. This has to end. No child needs 500 toys either!

I'm glad to see many practical alternatives to plastic packaging turning up in the UK. Some crisps, sweets, pasties, pies, cakes and breads are examples of what can be found plastic free at present. Some products can now be found packaged in natural substances such as potato waste and even seaweed. Other natural materials are being explored for the food industry, which needs to expand natural packaging to include all products in plastic or wrapped in plastic film.

We must end single use petrochemical plastic and use more bioplastic that is 100% 'Home compostable' and will break down harmlessly in a number of weeks. I stress that last requirement as only being compostable is not good enough for soil-health when it can take years for items to break down in regular landfill.

Chapter ten

Let's end single use plastic!

'What you do makes a difference, and you have to decide what kind of difference you want to make.' – Dr. Jane Goodall

We must put a halt to our voracious consumer habits if we want to purge single use plastic from the planet. Conversations about being Anti-plastic are a challenge but we must be fearless. So, make them loud. We need these conversations because the world hasn't woken up to the dangers of plastic yet. We must be the seed-planters for much needed change. I often sidle up to other customers in local supermarkets and talk about the horrors of plastic. Sometimes these little eco-chats bear fruit and at other times I get a blank look, but I will keep going.

We must break free from the all-powerful chemical and plastic industries that have held the world in thrall for far too long. Sadly, emerging countries only see plastic as a blessing and are going to be extremely hard to convince about its dangers. Our watery world won't allow Herself to be polluted without serious consequences and now Her waters are vomiting back our filth whenever there's a storm. She is sick but it's within our power to cure Her.

In 2020 Tesco UK warned 1,500 of their suppliers that packaging would be a deciding factor on which products were sold in its stores. This is what all supermarkets should do, threaten not to stock environmentally unfriendly items. We don't need to kill everything in sight in order to have a sustainable world! What we the consumer can do in the meantime is:

- Pester!
- Complain!
- Campaign!
- Vote with our wallets!
- Be brave loud voices for change!
- Remember: We don't have to buy what they're selling!

There is no excuse for single use!

We need People Power more than ever!

Our governments are snails when it comes to anything that will 'damage' their economy. So, write letters to company CEOs, email, tweet when you can, and generally persuade if they are not budging on plastic. Be polite. Flatter first and then go for the jugular!

Most corporations just accommodate themselves and their shareholders. They obviously don't give a toss about the planet. Find out which other brands they own and boycott **all** of them if you can. They will sit up and notice if we hit them in the pocket. Take down the names and details of the producers if you can. You can also email or message them on social media. Many producers and most of the main retailers will have a

Facebook page on which we can complain. Don't expect miracles but sometimes they do listen, and we must always give credit where credit is due to encourage them to do more good things. I do complain each time I see something mad and stupid, which, sadly, seems to be quite common now. Take the horrendous idea of putting glitter or flocking on living plants, which will rapidly suffocate them!

What we could say to companies who package in plastic:

Dear...
I really enjoy your product but I'm concerned that it still comes in plastic when there are alternatives available now and other producers are using them. Please assure me that your research and development department is working on more natural and sustainable packaging for your own products.
Kind regards

If you do receive a positive reply, thank them again for making the effort, kindness costs nothing:

Thank you for the courtesy of a reply. Most companies contacted just ignore customers' concerns. I'm pleased that you are working towards better packaging to help save our wonderful world.
Kind regards

The time for pussyfooting around plastic is over but ideally what we mustn't do is continually swap one type of plastic for another. Although, sometimes we do need to compromise. I hate any plastic with a fervour but we're in the early stages of ridding the planet of single use. Anything that contains a much smaller percentage of plastic and has a much longer life than single use has to be considered at this point in time. Ultimately, we need to wean ourselves off petrochemical plastic altogether.

The **Swaps** section next is not meant to be comprehensive, and couldn't possibly be. I'm so pleased to see new innovations turning up almost daily. It will certainly help you rethink plastic and find your own way through this minefield.

Researching these alternatives has given me a great deal of hope for the future.

I've tried to include many items, and issues around plastic, that concern us on a daily basis. It's in Alphabetical Order for easy reference and most of the more unusual items can be purchased in Plastic Free/Zero Waste shops or from that kind of online business.

Cleaning plastic free tips come with the name of the item to be cleaned. I have mentioned a few brand names but that's because they are one of a kind, clever/new Earth-friendly products we should be aware of. There is also a more detailed list at the back of the book.

I have a PDF of mostly UK online businesses and physical shops on my Yes, there are Alternatives to Plastic Group on Facebook. Ask to join and then go to Files when accepted. Anyone is most welcome to post anything plastic/Eco/planet/climate-related to this page.

We're all in this together!

If we want plastic-free products, including packaging, we should only buy from Plastic free/Zero waste businesses. They care!

If we also use Ecosia Search to find them online, they plant millions of trees in a large number of countries to help green and rewild Mother Earth. So, with a couple of clicks we've helped create a new forest somewhere too.

Trees are the lungs of the planet!

If we plant more trees around the globe, we will not only be regreening Earth, and improving our air-quality, but will create more shade in our rapidly heating towns and cities.

Swaps

Thinking planet first, not plastic

'Do the best you can until you know better. Then when you know better, do better.' – Maya Angelou

Before buying anything, we should ask these 8 important questions:

- Do I/my family/business really need it?
- Can I/a friend, family member, make it?
- Can I borrow it?
- Can I get it second hand?
- Could I hire it?
- Can I buy one to last a lifetime?
- Can I buy it plastic free?
- Am I being an Ethical Conscious Consumer to help the Planet?

If you spot the 'B Corporation' logo on a company website you can be sure that this business is doing all it can for the planet. The B stands for 'Benefit'. Certified companies are legally bound to consider any impact of their decisions on suppliers, workers, customers, community, and the environment. More of these companies are turning up every day and we must support them and buy from them if we want a cleaner and healthier world for all life.

When shopping for groceries, see what you can buy plastic free and then plan meals around what you can find. If you take a list of the items you want to buy plastic free to a shop or supermarket you will go nuts trawling the shelves to find them! Also use whatever you have in the kitchen already before you go and buy more food. Having little spare money has certainly made me appreciate this good habit.

I don't use anything toxic at all now. It can be done. I feel a lot healthier, and happier knowing that I'm not adding to the problem. There are so many alternatives that were common in the past and I hope people will realise that the problem is not only the plastic bottle/container but what is inside it.

Dip into the following **Swaps** section whenever you need a plastic-free alternative and plastic-free cleaning tip. I've wracked my brains and asked around to find as many everyday solutions as I can. As I mentioned at the start of the book, this is not a comprehensive list but it will help start the journey. I'm sure I've missed something but then Ecosia Search is a good place to start. Remember, they plant trees that help green and rewild the planet.

Swaps
No-one needs plastic when the world is full of beautiful natural sustainable materials

Four natural cleaning agents mentioned throughout this list:

White Vinegar or Apple Cider Vinegar can be used for general cleaning and for cleaning glass.
Lemon Juice cleans away grease.
Citric Acid can be used as a toilet cleaner.
Bicarbonate of Soda is a good general refreshener and cleaner. It is also good for safely and gently removing dirt, mould and algae from grouting.

A

Air freshener

We should avoid chemicals that make the house and furnishings 'smell nice'.

- Make a freshener using essential oils, in a metal or glass pump-bottle. There are plenty of good recipes online. Copper or brass plant mist sprayers would be ideal for the job and look attractive if left out in a room.
- Scent wood using essential oils of choice.
- Small wooden blocks are available to scent from craft shops and online businesses. They can be displayed in an attractive bowl too.

- Bicarbonate of Soda can be used to freshen a lot of things, including clothes in the wash, carpets, furnishings and fridges.
- Potpourri is an effective way of freshening a room.
- Buy an air-cleaning houseplant like a Rubber plant, Spider plant or Fern. Most plants indoors will clean the air and contribute to a healthier atmosphere. Choose ones with a long life.
- Open windows and doors on good days.

Athletic/leisurewear

This kind of clothing often contains Elastane, plastic fibre, to retain shape. Some companies who care about the planet and sustainability make their clothing from 100% Bamboo fibres, Hemp or real Wool. We should support those if we can.

Antibacterial handwashes

'The ten dirtiest things on the human body are the ten fingers. No matter how dirty your hands are, you can protect yourself with regards to indirect transmission and your hands. When you wash your hands, it isn't the soap that's killing the germs, but it's the rubbing back and forth and the action of washing that is removing the germs. The fewer organisms that you have left, the lower your risk of getting sick.' - Dr. Philip Tierno, professor of microbiology and pathology at NYU School of Medicine and NYU Langone Medical Center

Most antibacterial gels/washes contain Biocides that kill the good bacteria on our hands, necessary to keep us healthy, along with the bad. We really don't need them in our homes. We should be using bar soap and hot water. There are soap-bars to suit everyone now. Germs will not be transferred to another user because they're washed away during the hand-washing process. Soap-bars are also great for travelling as they're not subject to the 'liquids rule' at airports, and can be kept in a soap-box or tin.

African Black Soap is traditionally made by tribeswomen in Ghana. It consists of plant materials such as plantain, cocoa pods, palm tree leaves and shea tree bark, with no artificial ingredients, and must be kept dry between uses. This all-natural soap is reputed to be good for problem skin.

Artificial grass/Astro turf

We shouldn't be using it anywhere! It needs to be banned worldwide for the dangerous substance it is.

- It is neither child nor pet friendly!
- It's bad for wildlife, creating a dead zone.
- Herbicides and microbicides are used to kill anything nasty that can get onto it.
- The backing is manufactured using PFAS which then leaches out.
- It is treated with powerful chemical flame-retardants.
- It causes toxic runoff after rain.
- The microplastic that it sheds needs to be washed off the skin, as reported by sporting people.

It can cause severe skin irritation and get into our lungs.
- It's proved hard on athletes' bodies as it can cause injuries and rug-burn.
- It can get hotter than asphalt in the Sun.
- Concerns have been raised about Silicosis and elevated risks of Staph infection.
- We are breathing in any plastic microparticles released into the air!
- Sadly, can encourage laziness and a wildlife doesn't matter attitude.
- There is a huge problem in the USA with rolls upon rolls of artificial grass piling up in landfills as they became a popular alternative to the real grass lawns and sports fields that required regular and expensive upkeep.

We should be turning domestic lawns into wildflower patches/meadows for pollinators and other wildlife. We need more green roofs too.

Real trees, plants and grass give life. Plastic ones take it away!

At work

Take a healthy and plastic-free packed lunch if you can, to avoid all the frustrations with plastic packaging and those enticing eateries that will cheerfully drain a bank balance. You could take soup or a pasta dish in a flask. There are some great stainless-steel food-flasks available now. Keep jars and bottles of your favourite condiments at work to avoid all the sachets and pouches, etc.

Put a reusables kit in the drawer, cupboard or locker that includes the following:

- Plate
- Bowl
- Cup/mug
- Water-bottle
- Cutlery
- Chopsticks
- Napkin
- Shopping bag or tote for those lunchtime trips to the shops

Use what you already have at home and don't buy anything especially for the job if you don't need to.

B

Baby care products and clothes

- Organic baby care products will keep delicate skin hydrated, moisturised, and irritation-free.
- Bebeco has been selling natural products in the UK since 2004, and there are many other companies who specialise in all-natural mother and baby care.
- Use organic cotton, nappies/diapers, towels and clothing if you can and avoid anything synthetic.
- Source baby bottles in glass with a real rubber teat. Also, source rubber dummies/pacifiers and teething toys made of the same natural rubber from the Hevea Brasiliensis tree. We don't want precious babies ingesting any plastic!

Baby food in plastic pouches

- Avoid these plastic nightmares and look for good food sold in reusable/recyclable glass jars.
- Make your own with fresh fruit and vegetables – lots of great recipes online – to store in small glass jars in the fridge.

Backpacks

- Try to avoid backpacks with plastic fronts, straps and pockets. They will rip/fall apart more quickly.
- Check that zips are good and fastenings are secure.
- Avoid polyester ones made of recycled plastic bottles.
- Nylon ones won't help the plastic crisis either.
- Choose ones with a neutral design or no design at all so you are not persuaded to replace them when they go out of fashion.
- New or recycled canvas backpacks are strong and will last a long time and can be bought from camping/outdoor adventure stores and eco-businesses. Some of them are waxed to make them more waterproof and can be re-waxed later on if necessary.
- Before plastic, canvas was the go-to material for all well-made backpacks, and is still used by certain eco-companies.
- Vegetable-tanned-leather and beautiful Cork are sustainable options now as the world of plant/vegetable/fruit-leathers is growing fast.

Balloons

A huge dangerous and toxic problem in the environment! Say no to balloons altogether and not just releases!

Please don't inflate to celebrate any occasion. There is no such thing as an Earth-friendly balloon.

Any balloon is a potential sea life, wildlife and domestic animal killer. What goes up must come down. Many marine species such as seals, whales, dolphins, sharks, albatross, and turtles as well as land animals such as cows, horses, dogs, sheep, and birds have also been spooked by, hurt or killed by balloons and their trailing strings. If they ingest pieces of balloon, that block the digestive tract, it leaves the creature unable to take in food or nutrients. It then slowly starves to death with a stomach full of plastic. Balloon strings can entangle or strangle an animal preventing it from moving or eating and, if not found and rescued, it will die of exposure and starvation.

(Having just mentioned sharks, please avoid cosmetics that include **Squalene** which is found in the shark's liver. Sharks, which are already highly endangered from shark-finning, for Chinese/Asian Medicine, and pollution, are vital apex predators in our oceans. Some of the larger species with the biggest livers don't reproduce fast.)

Balloons with glitter or confetti in them are twice the danger of ordinary balloons. When popped the plastic contents explode everywhere. They can then be hard to find and pick up for disposal to prevent animals accidentally eating them with the grass. If you come across one of these, as I did in the countryside, don't pop it! Take it home and let it deflate over the week and then bin it whole. (What I should have done, even though I was careful to pop it when it was in a bin! I learnt the hard way!) As modern latex rubber used in some balloons also contains plastic it will never fully degrade in the environment. The original real latex rubber was biodegradable. Modern balloons don't biodegrade at all. Mylar novelty shiny metallic balloons could remain as litter for decades and are a constant danger to wildlife and sea life whether inflated or not. If these Helium-filled balloons get caught on power lines, because they are metal and therefore conductive, they will cause an Arc Flash and explode, causing power disruptions, fires and injuries!

Helium, used to inflate balloons, is a finite source that is rapidly running out. It needs to be used for medical equipment such as MRI scanners. Balloons used in releases are often coated in Hi-float (equivalent to PVA glue) to stop the helium gas from leaking out. More toxic pollution!

Turtles have mouths with backward-facing fleshy barbs that allow food in but not out. If they accidentally ingest any floating plastic film/bag, balloon or balloon-string it stays in their body and eventually kills them from starvation. Thousands of creatures a year are washing up with stomachs full of plastic, including balloons.

There must be worldwide bans on balloons of all types as the balloon-habit is far too ingrained now. The balloon lobby is another powerful billion-dollar industry. It's going to be some fight! The best way to be sure about balloons is to go without them. Please sign whenever you see a ban the balloon/balloon-release petition.

Water balloons

If you are ever reluctantly involved in a water-balloon-fight please ensure that all the pieces of burst balloon are removed and binned before they can enter the environment.

Reusable water-balloons

- These can be crotched in cotton and other natural materials from patterns online. They are rapidly becoming a popular and safe alternative for a good soaking with none of the attached danger to the local ecology. Please recommend them to your family and friends.
- Sponges can be a good alternative for a water-fight.
- You can get natural sea-sponges that are sustainably harvested so that regrowth is possible. They would be a lot softer too. Just keep them dry between bouts or they will go mouldy!

Alternatives to balloons for remembrance:

- Plant a tree or pay for one to be dedicated.
- Plant a gorgeous flowering bush/shrub, anywhere it is needed.

- Fruit bushes will benefit every form of wildlife with their wonderful flowers and tasty fruit.
- Plant some lovely bulbs somewhere wildlife can thrive.
- Plant some wildflowers, even a meadow.
- Do something special, like sharing a favourite meal that your dear one would love.
- Dedicate a bench to them on a seafront or in a park for all to enjoy.
- Buy a lovely birdbath for a garden or dedicate one in your own special space.
- Buy a beautiful statue for your outside space to always remind you of their love.
- Dedicate a small ornamental tree for your garden or outside space. Japanese Maples/Acers come in beautiful colours and shapes and are long-lasting when well cared for.
- Go out and do something that you used to enjoy doing together.
- Maybe do a memorial walk in the countryside with mutual friends, enjoying warm memories.

Alternatives to balloons for parties:

- Colourful homemade Chinese or Indian paper lanterns – instructions online or in books – make charming decorations for all sorts of locations and occasions.
- Recycled material or paper bunting is a cost-effective choice.
- Paper windmills - instructions online or in books – look lovely.
- Wool pom poms are becoming popular. Each year, sheep produce new fleece and will do so as

long as there is good grass for them to graze on. The shearing of fleeces helps the sheep keep cool, clean and parasite-free. This makes wool an excellent renewable natural fibre and a superb alternative to plastic. Wool should never have been abandoned in favour of plastic fibre. I'm so glad it is becoming the popular choice again. Sheep that are fed on good pasture, that takes Carbon dioxide from the atmosphere and deposits it in the soil and plant-roots, utilise carbon to grow their wool. All good for our dear Earth!

- Round paper and wire lanterns in various colours or animal-shapes, hung from the ceiling look very attractive.
- You could buy ordinary white ones, that are cheaper, and paint them/personalise them using dry-brushed watercolours or stuck on paper shapes.
- Some people like a number-balloon so they can take a picture of the birthday girl or boy surrounded by their presents. You could buy a collapsible wire and paper lantern and stick the number on it. It can then be used every year as the child grows.

Bath lily/pouf/puff

- The Loofa is a popular choice but you have to soak it first to soften it or it's scratchy on the skin. It's actually a plant, one of the Squash family, and can be grown in the garden in warmer climates or in a warm greenhouse.

- Cotton knitted bath lilies are available from artisan makers or can be hand-made. Instructions online.
- String exfoliators can be found in most chemists.
- Cotton and linen ones are available in some chemists/drugstores and plastic free/zero waste shops and online businesses.
- Natural materials need to dry out thoroughly between washes or they will become a black mouldy mess!

Bath mat

- Natural rubber ones that will biodegrade at the end of their lives are available.
- You will probably have to search online for the quirky or unusual versions that are popping up now.
- Keep them mould-free by hanging them up to dry after use.
- Some people recommend slatted wooden mats but they can't be used in smaller spaces.

Bath mitt

- Cotton, linen or Hemp bath mitt - one side is rough for exfoliating and the other is smooth, usually only packed with a piece of card.
- Konjac sponge – made of plant-root material and home compostable.
- Loofa – a plant. Soak before using or it will be a little scratchy.
- Cotton flannel – cheap and effective. (Flannel fan forever here!)

Before home composting, just wash through any natural materials with some clean fresh water to remove any soap residue. Natural soaps won't do any harm but they might encourage 'smelly' bacteria.

Bath toys

- Cork is waterproof and will not rot or degrade over time, and dries fully between bath-times. It's non-toxic and completely natural so won't harm any little chewers. Of course, it floats too for years of bath-time fun.
- Cork is fully sustainable and is harvested harmlessly from the Cork Oak which then regenerates its bark.
- Real rubber toys, painted with eco-paints, are available that will last a long time and can then be composted at the end of their life.

Bathroom sets

Can be found in:

- Glass
- Stainless-steel
- Ceramic
- Bamboo

Sometimes the dispenser top on the pump is plastic, but 95% plastic free is better than nothing.

Beach toys/Sand toy sets

- Metal buckets and spades are still made.

- Roguewavetoys.com make a fully compostable bucket and spade set by using 'bio-based material obtained from renewable resources.'

Bead-stringing/jewellery making

Silk thread, cotton cord and natural rubber elastic is available and we don't need to go down the plastic thread route every time we want to make jewellery such as necklaces and bracelets. Also, a lot of jewellery findings are made of natural materials, including silver, which are beautiful and still relatively cheap. Try to buy recycled silver whenever you can to tap into a no-waste economy.

Bean bag toys and chairs

There is no need to use Styrofoam beads when plant-based ones are available from Bigbeanbagcompany.com, who also make their own chairs with all-natural materials.

Bedding/Sheets

Don't be the human ingredient in your polyester/plastic-bedding sandwich!

We breathe in plastic fibres each day from around the house. Try to avoid the memory foam pillows and mattress-toppers and the polyester duvets and sheets for you and your loved ones, and be prepared to pay a great deal more for natural materials if you need to buy new.

Look for:

- Charity/thrift shop finds in cotton – just give them a good wash.
- Organic Cotton – sustainably sourced if possible.
- Silk or silk and cotton mixes.
- Eucalyptus.
- Hemp – a fast growing and sustainable plant. (I'm all for it as long as it doesn't create another monoculture. Mother Earth needs diversity to thrive.)
- Bamboo - has antibacterial and antimicrobial properties just like most wood has.
- Duck down duvets and pillows – just make sure they're cruelty free and the feathers are harvested during the ducks' natural moulting season.
- Wool duvets and pillows – come at a hefty price but are healthier than polyester ones.
- Camel hair duvets – come at a very hefty price and are made from the soft hair that is shed naturally each year. It takes a lot of camel just to make one duvet!
- Long-lasting latex rubber and cotton or wool mattress toppers can replace memory foam This popular plastic material usually doesn't last long, stains badly, and ends up in landfill or incineration.

Organic cotton Pillows can be found, filled with other natural materials such as:

- Organic wool
- Tencel – made of plant fibre
- Millet

- Buckwheat
- Spelt

Search out companies that make complete beds out of natural materials, mattresses, toppers, bedding and beds for babies, children and adults.

In the handmade and bespoke market, you can find Mohair, Cashmere, Bamboo, Cotton and Natural Latex in their ranges.

Fabric made from organic Hemp can be as soft as silk, drape as beautifully as linen and be as snug as any fleece. It is as versatile as any cotton fabric and can be made as strong and long-lasting as canvas.

Many believe it is the material of the future that will help save the planet.

Bird food

It is vitally important that we feed our garden birds all year round because of the loss of natural habitat. Most food for our treasured feathered friends is, sadly, still available in plastic but some companies are changing that as soon as they can. We can always write to others to encourage them to change too.

If you buy bird food loose at a pet store, take your own container/s to avoid the use of plastic bags. It is an easy habit to get into.

Use plastic-free bird-feeders if you can. There are good strong ones constructed of stainless-steel or wood and slate. You could be creative and make your own from old metal kettles, ceramic cups and teapots hung in trees and bushes. You can also use old kettles and teapots to make nests for small birds. Numerous solutions and projects can be found online and in specialist books. The RSPB in the UK is a fount of knowledge on all bird-related topics.

Biscuits

Try to buy tins or cardboard boxes of biscuits if you can or make your own, with the ingredients sourced plastic free if possible.

Danish Butter Cookies are always available in tins at Christmas, and in paper. The lovely tins make nice little lunchboxes afterwards or you can reuse them to store other biscuits and cakes. They are good for storing umpteen little bits and pieces throughout the house and garden. Every household when I was growing up kept biscuit/sweet tins for storage. We kept our Christmas decorations and lights in large tins in the loft.

Bleach

Bleach is the generic name for any chemical product which is used industrially and domestically to clean and remove stains. It often refers, specifically, to a dilute solution of Sodium Hypochlorite, a weaker version of Chlorine. Bleaches can react with many organic substances so they can weaken or damage natural materials. For the same reason, ingestion of the products, breathing in the fumes, or contact with skin or eyes can cause serious bodily harm.

Keep them out of reach of little hands and vulnerable people!

Alternatives to Chlorine Bleach

- Bicarbonate of Soda can safely be used to clean and sanitise toilets and so can Citric Acid. Both can be sourced in cardboard in the UK.
- Oxygen bleach is another safe alternative.

Body-boards

Just when you think you've seen all the plastic pollution there could possibly be, there's even more in the form of these blighters that seem to be everywhere on every beach worldwide!

Cheap body boards are filled with Polystyrene balls that escape into the water when broken and can be ingested by sea life. Another nightmare product! It's in microparticles in seconds in some cases.

These boards need to be banned, along with flimsy plastic sledges that are often discarded on the snow after one use! We should be buying good quality boards that will last years of watery or snowy fun.

Bottle bags/carriers

Avoid the plastic covered paper shiny ones with drawstring bags made from:

- Pure organic cotton
- Hessian
- Hemp

Make one using a natural, strong, fabric. You can also make one with a scarf or a large square of fabric using the ancient and artistic Japanese technique of Furoshiki fabric tying - instructions online and in books. Second-hand scarves and even small cotton tablecloths can be used, becoming a charming bonus gift.

Bottled drinks in plastic

A great number of these will have artificial sweeteners in them, such as Aspartame, that are not good for a healthy body, especially a child's growing one. Always read labels and try to source natural fruit juices in glass bottles. A lot can be found on supermarket shelves now to avoid the plastic and chemicals. Buy fresh fruit to squeeze, and bottle your own fruit juices when you can.

Bottled water in plastic

It takes 6 times more water to produce one bottle than to fill it!

We must boycott greedy bottled water companies. They are draining our world rivers, destroying Mankind and precious Earth! They can't be allowed to do this anymore! The age of bottled water has to end and we need world laws against draining rivers for profit.

Alternatives:

- A stainless-steel water-bottle is so much healthier for us than a hard plastic one and will last years.

- Use your own tap-water if it is good and watch out for refilling stations when out and about.
- You can buy a water-filter if your tap-water isn't good and decant the fresher water into your reusable bottle.
- A Bamboo bottle – buy a good one that uses Earth-friendly glues.
- A glass bottle is also an alternative but may not be as practical in certain situations or for children.

You can download the **Refill App** in the UK which will tell you where you can fill up for free. Look for the 'Waterdrop' sign that is appearing all over the country now as we wake up to plastic-water-bottle pollution.

It is not advisable, according to science, to reuse any single use plastic water-bottle because of the toxic Dioxins that leach from the plastic over time, especially in a warm environment. This particular sort of plastic is not designed to be used more than once. **Be safe and not sorry!**

Boxes for storage

We all have plastic boxes that will last us a few more decades and hopefully within that time-frame plastic will be dying out and we will have safer ways of disposing of all existing plastic.

Look for natural alternatives for boxes in:

- Wood – antique or from sustainable sources.
- Metal – Ant-proof travelling trunks are a great option and look cool too.

- Wicker
- Seagrass
- Cardboard - for storage in drier atmospheres.

Bras

Good luck finding any made of 100% natural materials! Most are now made of mixed fibres with added Elastane, plastic. The Bamboo-fibre step-in/pull up ones usually also contain Elastane but very little compared to the more traditional bras. One of the ways to source all-natural cotton bras is to buy from artisan makers and be prepared to pay a premium.

We must push for an end to petrochemical plastic in clothing that is no good for us or the environment. We must wear 100% natural and sustainable fibres once more if we want healthy bodies living on a healthy planet.

Breakfast cereals

We all know how difficult it is to buy cereals in anything but a plastic pouch inside a cardboard box, or in a completely plastic pouch. Cereals used to come in only paper and cardboard when I was growing up. Muesli can still be found in cardboard boxes, and porridge oats/oatmeal in paper, in a few major supermarkets or in organic or zero waste/plastic free shops and online businesses.

Our 2 popular budget supermarkets do oats in recyclable paper packets at reasonable prices. Having been brought up in a Scottish household, I know the benefits of a quick-to-make lovely bowl of steaming hot healthy porridge oats on a cold day!

Plasticfreepantry.co.uk is a great go-to website for kitchen cupboard staples that come 100% plastic free, as well as their own coffee.

Brooms, brushes and dustpans

Look for a broom made with a wooden handle and plant fibre bristles.

Metal dustpans and wooden brushes made with plant fibre bristles are still around. They used to be common in the old Ironmonger shops but now you may have to search online. I do have a set myself, so they are out there.

Bulk-buying

This is one sure way of cutting out a lot of plastic. Bulk/zero waste/plastic free shops are popping up in many UK towns and cities now and will soon be the norm on our highstreets, and might even save them. They will all be a bit different but **Organisation** is the key to bulk-shopping.

- Collect jars with screw-on or locking lids of various sizes for seeds, nuts, pulses, cereals, spices.
- Use any suitable containers.
- Some will probably be available to buy in the shops themselves but save yourself the trouble and money and take any you already have in the house, even plastic ones.

- Take your own reusable produce-bags and shopping bags, and a box to load into the car if a driver.
- Take someone with strong arms if you need them!
- You may need a funnel, metal is best, to transfer nuts, pulses, etc., to jars from bags, if used, when you get home.

If you collect a lot of reusable bags, often given out at conferences, fairs and festivals, you can pass them on to zero waste and plastic free shops for when customers have forgotten their own. They will really appreciate the thoughtful gifts.

Bunches of flowers

These usually come in plastic.

Look for real Cellophane that is biodegradable, made of plant material. Some florists stock it as well as use it for their arrangements, or use paper. Just ask. Also look for flowers that can be wrapped in some sort of natural material like Jute or Hessian.

Buttons

In the past buttons would have been made of metal, horn, bone, antler, leather or shell all of which is still available. It's sad to see so many are now made of plastic imitating natural materials.

Cut off buttons from old clothing and keep in a Button jar or box/tin.

Look for these natural buttons as well:

- Wood
- Coconut shell
- Cotton
- Tagua – a nut that is known as vegetable ivory and takes colour well to make very attractive buttons.

C

Cakes

- Buy from a local bakers or market to avoid the plastic wrapping or clamshell packets, and take your own bag.
- Frozen pies and cheesecakes are often plastic free and just in a metal tray and cardboard box in the freezer section of supermarkets. Watch out for frozen cakes, though, as they often have a plastic collar or are completely in plastic.
- Make your own! There are fantastic recipes for different breads, cakes and meals online and in the usual cookbooks. You may not be able to go entirely plastic free but any reduction is good. No doubt the taste will be far superior to bought ones anyway.

Candles

Paraffin wax candles, derived from petroleum, release toxic air-pollutants when lit.

Beeswax candles can be a natural and healthy alternative and usually come in attractive designs from artisan makers. They make lovely thoughtful gifts.

Soy wax candles burn longer than paraffin candles and are:

- Plant based/Vegan
- Healthy
- Toxin free
- Naturally scented
- Sustainable
- Biodegradable

Car fresheners

Avoid the hangers, that are usually plastic, and put a few drops of your favourite essential oil on a wooden peg and peg it somewhere. I wouldn't advise using Tea Tree Oil with its powerful odour but there are many more that are suitable. Find a refreshing uplifting one.

Carpets/mats/rugs/runners and flooring

So many carpets are made of mixed plastic material. We are breathing this in every day in the form of the microfibres they shed constantly and when being cleaned or walked on. If we want a healthier planet for all we must use natural carpeting and flooring whenever we can.

There is a huge choice of natural carpets. Among them:

- Wool
- Cotton

- Seagrass – hard wearing.
- Bamboo
- Jute fibre – made from the Jute plant.

For flooring there are these natural choices:

- Reclaimed hardwood
- Cork – warm underfoot as well as being fire retardant and water resistant.
- Natural Linoleum/Lino made from linseed oil, cork dust, tree resins, wood flour, pigments and ground limestone. Lino is also easy to clean, fire retardant, water resistant and long lasting.
- Recycled glass tiles for kitchen and bathroom floors are becoming popular as glass bottles are given a good second life.

Carpet freshener - homemade

- Fill a recycled glass jar with Bicarbonate of Soda.
- Add your preferred essential oil, about 20 drops.
- Screw the lid back on.
- Give it a shake and then leave it sealed overnight for the oils to soak into the Soda.
- Pierce holes in the jar-lid and sprinkle the Soda directly onto the carpet.
- Leave it for about 15 minutes and then vacuum it off.

Bicarbonate of Soda can be sourced in a cardboard packet in the UK.

Cat litter

Not plastic but usually comes in plastic pouches and is still an environmental nightmare/health hazard for disposal. Some people even put it down the loo!

It should never be flushed whatever it is made of as it will clog the pipes!

There are alternatives to the hard clay granules that end in landfills all over the globe and take a long time to break down.

Bio Catolet is lightweight, non-clumping, and 100% recycled paper that is pressed to form absorbent granules. It's dust free, all natural 100% biodegradable, insect repellent and prevents bacterial growth in the tray. It is kind to cats' paws too. It can be easily disposed of via the household waste bin, compost heap or can be incinerated. It's also manufactured without causing any pollution to the environment. Sounds like a win all round to me!

Other biodegradable cat litters are:

- Pine wood pellets
- Clumping sawdust
- Brazilian Cassava plant
- Corn
- Wheat waste - which would normally be burned or put into landfill.
- Walnuts
- Barley

- Dried orange peel
- Soy pulp made from waste from the industry.

Celebration decorations – Easter, Halloween, Christmas, Birthdays, Weddings, and other festivals and special/holy days

It makes any 'Eco Worrier' sick to see all the 'shiny' decorations. I try not to go to the shops that sell all that plastic pre-landfill crap just to keep sane. Try to avoid all the plastic tat for decorations and presents/gift bags and make or source natural sustainable ones. Be creative! Also let family and friends know that you want to be more mindful about gift giving to help save the environment from plastic and excessive waste. Suggest alternatives and give them a copy of this book. Your family and friends will be much more likely to get onboard if you let them know how you feel and what you are going to do to achieve your goal.

- Natural decorations can be made using dried citrus fruit skins. Make shapes with metal cookie cutters and string them as a garland or hanging decorations. They smell divine too!
- Needlefelt wool decorations can be bought from specialist makers, who may also offer a bespoke service, and some online stores.
- Don't forget that lovely paper snowflakes, snowmen, and lanterns are easy to make too. There are some fantastic designs on YouTube.

Buy second-hand costumes, make or swap. If you already have something that will do the job, use that if it is not harming you, your family or the environment. Lots of ideas online.

We should gift experiences, and not things, that will be remembered forever. We could take family and friends to one of our wonderful world-renowned UK wildlife conservation parks that are working daily to reintroduce endangered species to their natural habitat all around the world. They need all the support we can give them.

If you really want to give gifts, buy those that help wildlife and the environment too, such as an insect or bee hotel, a bat box, bird box, hedgehog home or plants, trees and wildflower seeds. You could give the gift of adopting an endangered species or supporting an environmental cause. Plantable wildflower cards and gift tags are also becoming popular thoughtful eco-gifts throughout the year. Just remember to plant them!

Avoid giving plastic Gift Cards as they have a huge environmental impact around the globe. Tonnes end up in landfills each year! Many gift cards are made of PVC plastic which is hard to recycle and isn't accepted by most recycling systems, and not by Councils here in the UK.

Look for retailers that:

- Use E-cards that can be printed out or redeemed online or from a phone.
- Offer cardboard cards.
- Use cards that can just be printed out.

- Can reload one of your own spent cards and pass it on to someone else. Not ideal but better than putting another one out there.

Celebration cards with glitter on them or shiny plastic surfaces and plastic decorations like ribbons and bows, etc., are not good for the environment. Go for plain printed carboard instead without the plastic pocket, or a real Cellophane one that is made of plant material that can be composted. Check the label.

Glitter is made of Polyethylene, the same substance that's in plastic bags, and gets everywhere.

Bioglitter/Ecoglitter still ends up as litter and can cause considerable harm if it gets into waterways as it takes a long time to break down.

The best alternative is to avoid glitter, in all its forms, altogether.

Securing parcels with biodegradable Japanese Washi Tape that comes in all sorts of attractive patterns is a good alternative to any plastic tape.

It is:

- Compostable or recyclable.
- Plastic free.
- Vegan – made of plant materials.
- Made with natural rubber-based adhesive.

Or you could just use paper tape that is available in some stationers, plastic free shops and online businesses.

Many designs available now. Maybe you could draw your own designs on plain paper tape for a bit of personalised fun.

Cellophane

This is all right to use as long as it is real Cellophane. Real Cellophane, made of Cellulose, a plant material, is compostable and was one of the first plastics before scientists started tinkering with them, making them almost indestructible.

Chewing-gum

The regular sort contains plastic and a host of chemicals including Aspartame, an artificial sweetener found in most soft drinks to replace expensive sugar. Not good for maintaining a healthy body.

Gum that is flushed down the loo or tossed carelessly in the street finds its way into waterways to poison river-systems and oceans.

Natural chewing-gum is available in many flavours now to suit all tastes. It won't harm the environment but must be disposed of in a bin, as it could trap wildlife and is still very sticky litter in warm environments. Everyone has had chewing-gum on their shoe at some point!

Chocolate bars

Bars in paper/card and foil are readily available once more.

Some UK supermarkets' own brands are now completely plastic free. Always check, though, because some chocolate bars are still in plastic in disguise as paper/card and foil.

There are many plastic-free Artisan Bars too, with exciting flavours, but be prepared to pay a lot more.

Chocolate spread

This usually comes in plastic, and often includes palm oil. Look for the ones in glass jars and check the ingredients.

There is more information about palm oil in **Washing up/dish washing liquid.**

Chopping boards – usually made of Polypropylene plastic

Alternatives are:

- Wood – naturally antibacterial.
- Bamboo – also naturally antibacterial.
- Tempered glass – easy cleaning.

Christmas – the season of plastic tat!

There are so many Christmas/celebration factories worldwide continually churning out more and more plastic decorations and plastic Christmas trees. The environmental damage must be enormous!

We must try to avoid buying any glittery and shiny plastic and use natural materials to decorate the house and presents. Don't forget that most Christmas lights are plastic too.

Felted real wool pom-pom garlands and cotton or paper bunting is pretty and would replace the plastic chains and tinsel, and look really different.

Paper decorations were all we had when I was growing up, and very charming they were too strung across the ceiling from the light rose. It was a lot of fun 'opening' the lovely tissue-paper Christmas bells and baubles and securing them with metal paperclips. They might have been paper but they were well made and lasted for years.

YouTube has some great videos on how to make paper decorations to replace all the usual plastic rubbish.

One of my favourite, no-real-skill-required, decorations is made by cutting finger-sized pieces of branches, on the slant at the top and flat at the base, a little thicker than a finger, and painting them with plastic-free paints to look like characters with a Santa hat, and long white beard. The wonkier, the better! A wooden bead then makes the nose between the paint-dot eyes.

They're called 'Little Wooden Men' in Germany where they originally come from. I arrange mine in threes on the mantelpiece so they look like they're singing or have been out carousing!

Christmas crackers – avoid all the shiny ones. They are usually plastic and contain plastic gifts.

- It's good to see that some UK supermarket chains and organisations have introduced completely plastic-free crackers.

- There are a lot of Make Your Own Crackers videos online and some great books about how to create a natural Christmas.

Christmas gift tags

These can be made using old cards and string. Cut around interesting shapes or characters to make attractive or quirky tags. This is the way we used to do it when I was a child.

Christmas jumpers/sweaters

Most are made of plastic fibres. How about getting the family involved in a project to make designs to sew on and reuse any jumper? You can even safety-pin them on for easy removal, and a rather cool retro Punk look.

Christmas trees

- Go real tree to avoid the plastic ones if buying new.
- Rent one that will go back to the farm after Christmas. Tree-rental/hire companies send their trees in biodegradable netting, in a sustainable pot, and pick them up in January to be replanted. The sustainable way of doing Christmas is catching on fast here in the UK.
- Real miniature table-trees are normally only good for the season, and please avoid any of the flocked ones as flocking is plastic. How anyone can do that to a living being is beyond me! The poor little trees will suffocate and die in a matter of days.

- Some shops are selling gold or silver sprayed living plants and young trees. It doesn't matter whether it's safe paint or not. Wonderful Nature doesn't need spray-painting or flocking! If you see this, please email the company, call it out on social media, or complain on the spot. We need to stop this!
- If you still have an old plastic/fibre optic Christmas tree, use it until it 'dies'. Dispose of it responsibly, and replace it with something natural. The same goes for any baubles and decorations.
- Most UK Councils will take real cut Christmas trees for shredding after Christmas and some zoos will take them as food for camels and goats and stimulating toys for larger animals.
- Real potted trees can be planted in the garden if you have the room or kept in the pot and put in a sheltered spot. Never let them get too dry, especially in hotter months.

Alternatives to the traditional tree:

- Trees made of different sizes of driftwood, tied together with rope, like a ladder.
- A life-size Christmas Tree material wall-hanging. There are some lovely ones now if you don't have room for a tree.
- If you're an artist, like me, buy a big canvas and paint one or a magical snowy scene that you can use every year.
- Make a minimal one from an old cardboard box or some lovely recycled wood – instructions online.

Cigarettes

Apart from being really bad for the health of the smoker and those around them, they are one of the **major plastic polluters on the planet and are dropped carelessly by the trillions.** They end up in our global waterways and oceans where they will poison fish and wildlife for decades.

Cigarette filters

These are made from cellulose acetate, a form of plastic which takes 12 years to degrade into microplastics in nature, never going away.

Plastic disposable cigarette lighters

The Spanish islands of Majorca, Menorca and Ibiza banned plastic lighters in 2020. They should be banned worldwide. In 2017 it was estimated that if all the plastic lighters that washed up on beaches were stacked on top of each other, the pile would be 10 times higher than the Eiffel Tower. Millions more have been thrown away on land or found their way into rivers, seas and oceans since then.

There's always a plastic lighter somewhere in my front garden as I live right next door to a Post Office/Shop where they're sold. Of course, people have to pitch out the old lighter after they've bought a new one. Or they drop it into my garden before going into the shop for a new one and a packet of ciggies. Then on leaving, they rip off the plastic and throw that into the garden as well. No wonder I have to clear my garden at least twice a day!

If you are going to smoke, please invest in a forever lighter made of stainless-steel or use matches that at least biodegrade.

Vaping also involves a lot of plastic and is proving to be just as destructive to bodily functions as smoking for both the smoker and from second hand exposure. Flavoured vaping has been known to clog the lungs and impede oxygen intake.

My advice, as I have lost so many loved ones to smoking-related diseases, is to save yourself, your loved ones and Earth, and give up smoking in any way altogether. Why let your hard-earned money go up in smoke?!

Cleaning sprays for surfaces

We don't need them!

A study has found that using cleaning products on a daily basis is just as bad for you as smoking a pack of cigarettes a day. As we women traditionally do the cleaning, we're the ones most at risk of chemical poisoning.

A steamer, cotton dish cloth or old tee-shirt rag, using only hot water from the tap, will do the job just as well. This is how we did it before nasty chemicals in spray-bottles. Bonus: free and toxic chemical free and you can breathe easily while doing it.

Another option is a Swedish sponge cloth made of plant material and cotton that is naturally antibacterial. These are becoming popular in the UK now and have been used for decades in Sweden. They can be thrown into the compost when worn out because they are used only with water.

You can also buy sponges, made in the UK, formed of wood fibre that are eco-friendly, as they go through a less toxic manufacturing process, biodegradable and home compostable. Others are made with string on one side and wood pulp on the other.

How to look after natural fibre sponges:

- Keep them dry between uses.
- Sterilize by soaking for a few minutes in boiled water.
- Microwave damp for a few seconds.
- Put them in the dishwasher with a drying cycle.

A recent innovation from green suppliers is a natural Concentrated Cleaning Pod that when put in a reused bottle with tap-water dissolves to make a more dilute and Earth-friendly cleaner. Some supermarkets have started to stock something similar but check that the ingredients are all natural and plastic free before buying.

- A 'green' way to clean taps, sinks or any stainless steel is to take a slice of cucumber and rub it onto the surface you want clean. Not only will it remove tarnish and bring back the shine but won't leave streaks.
- You can also use the outside skin of the cucumber to rub off any crayons and markers that the children have used to 'artistically' decorate your home.

Clingfilm/Plastic wrap

Sadly, it has become the go-to way of preserving food to comply with health and safety in the food industry but that doesn't mean we need to use it in our homes.

We have a number of choices:

- Use a plate or saucer on leftovers in the fridge.
- Buy some small glass casserole dishes, often found in charity/thrift shops.
- Reuse glass jars for small amounts.
- Beeswax wraps can be used when out and about.
- Reuse old paper bags, cereal-packet insides, and Tupperware tubs, etc.
- If you already have something suitable for storage, use it.

Clips for open snack bags, etc.

- Try to avoid the plastic snack bags and go for nibbles in paper or home compostable packets if you can.
- Wooden clothes-pegs can be used to close opened snack and cereal bags, and they can also go in the freezer, without breaking from the cold, for opened freezer products as well.
- Bubbe Clip make clever sustainable reusable wooden bag-clips to keep food fresh and reduce food waste.

Clothing

According to the charity WRAP UK (Waste & Resources Action Programme) that works to achieve a circular economy through helping to reduce waste, develop sustainability and use resources in an efficient way, the value of unused clothing in our wardrobes has been estimated at £30 billion. Also, an estimated £140 million worth of clothing goes to landfill in the UK each year. This is absolutely disgraceful and has to stop for the sake of the planet. We must avoid any plastic fibres for clothing and go for natural materials. We must also buy to last and hand down, swap, or give to charity any clothes we've outgrown or no longer need. Creating a Capsule Wardrobe is becoming popular. Keeping the clothing, shoes and accessories we need and regularly enjoy wearing to a minimum.

We could start with a Wardrobe Audit and go through cupboards and drawers, and also check on things put away in bags and boxes.

We can ask:

- How do I feel about this item?
- Is it comfortable?
- Am I going to wear it again?
- Do I have other things to wear it with/accessories for it?

One idea to make choosing the ones to keep a little easier:

- Turn all the clothes on hangers the wrong way around, with the hook facing out, on the rail of your wardrobe/closet.
- As you wear them, replace them the right way around, hook facing in, to see what you regularly like to wear/have to wear for work.

- Sell good little-worn items on social media or to a quality second-hand clothes shop.
- Swap any no longer worn with friends or at a clothes swap event, or rare jumble sale.
- Donate to charity.
- Repurpose those you can yourself.
- Give to friends to repurpose.

The Founder of the ethical fashion range Point Off View, Marina Testino, wore the same suit for a month to shine a light on our over dependence on fashion. She concluded through the experiment that it was liberating not having to choose different outfits each morning, and that the only person who really cares what you wear is **you**.

I agree. **Money** has always fuelled fashion. Try having no spare money for a while and then see what you wear. Buying second-hand is an easy habit to get into. Give it a try for the planet.

Most people love their body-hugging jeans but stretchy jeans contain Elastane which is a plastic fibre added to maintain shape. There are a few companies trying to cut this out of their production line but if you want truly plastic free jeans the advice from the eco fashion gurus is to shop vintage.

There is an image of the plastic left in jeans after the cotton is dissolved doing the rounds of social media at present. It is truly frightening to see that there is a dense web of plastic left and that it can be clearly seen that the clothing had been a pair of jeans!

Avoid the Synthetics!

What should concern us is that anything 'fake' or 'faux' or, sadly, Vegan is usually plastic. Also look out for natural clothing materials that won't shed microfibres in the wash and will last for years instead of months.

I love the idea of having a **Future Fashion Drawer** for our children so we don't waste good quality long-lasting clothes just because we have outgrown them in some way. Prato, a town in Italy, has gone one step further and is transforming the fashion industry by collecting and sorting natural materials from used clothing, wool and cotton, and turning it back into usable fibres that are now in demand from designers. They started doing this years ago because no-one there could afford new clothing. This should be the way things are done from now on to avoid the appalling worldwide waste from fast fashion.

Alternatives to synthetics:
Wool - there are two certifications:

Responsible Wool Standard (RWS) is a voluntary global standard where sheep farms are audited annually to ensure that their practices meet the standard of welfare of animals, and the environment they are raised in.

The ZQ Grower Standard demands sheep be offered these Five Freedoms:

- From thirst
- To live naturally
- Free from discomfort
- Free from distress
- Free from disease

Again, the standard addresses not only animal welfare but environmental sustainability, the quality of the fleece, traceability, and social responsibility.

Other alternatives:

- Alpaca wool.
- Cashmere – originally produced from the soft hair of the Kashmir goat in India but now produced in China and Mongolia.
- Linen – made of the Flax plant.
- Hemp - plant fibre. (So much that is made of plastic now could be made of Hemp in future.)
- Cotton – but try to buy organic and sustainably farmed.
- Corduroy cotton for jackets, trousers, skirts and pinafore dresses is becoming popular again.
- Bamboo – fastest growing grass on the planet that absorbs five times more carbon than hardwood trees. It needs half the land cotton does to produce the same amount of fibre and it doesn't require irrigation or the use of pesticides.
- Ramie - the fibres come from the stem of a nettle plant called China Grass (Boehmeria nivea).
- Silk - a beautifully cool natural fabric but please check how it is produced. Most silk producers kill the developing moths in the cocoons by boiling

and adding chemicals. There are Moth-safe silks now, known as 'Peace Silk' or 'Ahimsa Silk', where the Silk moth cocoons are unravelled harmlessly, enabling the moths to live out their natural pollinating lives.

Fabrics made of wood pulp or Cellulose, plant material, and are biodegradable include:

- Viscose – can be made to resemble many other fabrics.
- Rayon – can be produced to imitate the feel and texture of silk, wool, cotton, and linen.
- Tencel – also known as Lyocell – is made from fast-growing Eucalyptus in an award-winning closed loop process that safely recovers any solutions and emissions.

Plastic clothing tags need to be banned in amongst the single use items. Most have to be cut off and are already reduced in size before they hit the soil. As charity/thrift shops use them constantly too, I think we need to have conversations with them about this. It's great to see that some companies realise the dangers and have found natural sustainable alternatives like recycled card and string.

Please write to companies that use plastic tags if you can and also avoid buying new clothes that have them attached if possible. This won't be easy but we do need to start complaining loudly about these 'silent' polluters.

Coat hangers

Flocked and anti-slip are made of plastic, flocked with plastic, break quickly and can't be recycled. Sustainable ones are around.

Instead use:

- Metal
- Wood - Cedarwood deters moths
- Bamboo
- Recycled card
- Wood pulp

Tip:
Wind a couple of elastic-bands around the ends to make them non-slip. Use the ones your Postie drops!

Coffee capsules

Mostly plastic and terrible global polluters!

You can get stainless-steel forever ones now that will fit most coffee machines.

Coffee filters

A reusable food-grade stainless-steel one will last a lifetime.

A few producers use thin aluminium foil capsules that can be recycled with metal. Collect them and then crush them into a fist-sized ball first to go through recycling machines.

Combs

- Bamboo – strong and long-lasting.
- Sandalwood – has a lovely feel and smell and is a delight to use as the combs made from it are handmade and beautiful. Any sustainable wooden combs make lovely presents. Other woods are available now as the world wakes up to the nightmare of plastic but make sure they come from a sustainable source if you can.

Commercial coffee cups

Millions of trees around the planet are chopped down to make them. Millions of litres of water are used in single use coffee cup production so that we can use them for 15 minutes and then mindlessly throw them away. They are also lined with plastic which makes them non-recyclable. Biodegradable cups are also not a good solution to the problem as they rarely end up being composted correctly and do not break down safely in ocean environments. Coffee shops throughout the world now offer glass jars to rental mugs and BYO (Bring Your Own) cup policies/discounts.

The change is happening but the only real solution is to take your own cup when out and about, and there are lots of different reusable ones available:

- Stainless-steel – lasts forever if well looked after and washes like a dream. Camping/outdoor adventure stores stock good cheap cups for everyday use.

- Enamel – from camping stores.
- Bamboo – get a good one as some of the cheaper ones are glued with less than planet-friendly glues.
- Some are even collapsible, made of Silicone, or telescopic, usually stainless-steel, so you can pop them in a bag or even a pocket.
- Rice husks - a natural by-product of harvesting rice.
- Coffee grounds – another natural by-product.

Silicone is made from a polymer mixed with oxygen and other elements. It is more like rubber than plastic. Its benefits are that it's pliable/bendable and can take cold and heat, but it is made using fossil fuels. There are also concerns that highly-heated Silicone products can leach undesirable compounds called Siloxanes. Personally, I avoid it altogether now I know more about it, although I did use some in the past.

Composters/Bokashi bins and Worm bins

All three do a grand job in composting organic waste. No more nasty plastic bags full of stinking rubbish, or icky kitchen bins!

You can get compost bins of varying sizes made of:

- Stainless-steel
- Bamboo
- Wood
- Old pallets – DIY instructions online

If you have outdoor space, a Composter or Worm bin is best. If you don't have outdoor space but room in your kitchen, then a worktop/countertop Bokashi bin might be the best option. You will need to buy Bokashi bran to start off the composting with this method. You can always give the ripe compost to a gardener or to an allotment if you don't have your own outside space. It will be highly appreciated!

With a Worm bin, you will need to buy some Tiger worms to start you off but some come with the bin. They will quickly have families if you keep them happy and well fed with kitchen scraps. You end up with lovely rich compost and Worm liquor plant-feed, drawn off with the tap. I've had one for years. The little wrigglers do a splendid job and I am very fond of them! I highly recommend *Composting with Worms, why waste your waste?* by George Pilkington. Everything you need to know about keeping your worms happy in a healthy bin is mentioned in an entertaining way.

Lots of advice, and videos, about these home waste management methods can be found online and in specialist books.

Note: Before composting/adding to a wormery/Bokashi bin or putting in a food collection bin, remove stickers from peels/rinds just in case they are plastic. Organic material that still contains these labels is likely to be rejected as contaminated, and also won't be good for any home composting.

Computers/Laptops, etc.

At present, after the hard drive is wiped/destroyed for personal security, old ones have to be recycled with electrical waste at the tip, but when you buy a new PC or laptop in the UK, the retailer is legally obliged to:

- Safely and responsibly dispose of your old one when you buy a new version.
- Set up alternative free disposal.

Check to see if there is a local charity that will take your old computer/s to wipe free of charge, refurbish and give to those in need, or dispose of in an environmentally-responsible manner.

Confetti

Most is plastic, especially the shiny sort. Use leaves instead, dried or Autumn/Fall leaves are best, and a hole- punch. You can get some attractive/quirky hole-punches now to add to the fun. Dry out any leaves on a tray or in a just used oven before punching and then store them in a glass jar for that special occasion. Just ensure those you use are neither too wet nor too dry.

Cooking

- It is inadvisable to cook/microwave food in any form of plastic, trays and film. For years it was advertised as 'safe' but new studies show that plastic can leach into food when heated.
- We should avoid an excess of tinned food that's usually lined in a thin layer of BPA, Bisphenol-A, a hormone-harming chemical, and eat more fresh produce when we can.

- We can use sauces and foods in glass if we don't make our own.

It is good to see that some companies aren't waiting to be regulated and are finding a non-toxic alternative to BPA because it is the right thing to do for us and the planet. We must be aware, though, that some of the many **chemical alternatives** to BPA are even more toxic than BPA itself!

Stainless-steel, iron, ceramic, earthenware or glass are the safer options when it comes to cookware as the coated ones contain Poly and Perfluoroalkyl substances (PFAS) - used to create the non-stick, stain-resistant, water-repellent surfaces, and are highly persistent both in the body and environment. They are a likely carcinogen and have been shown to cause cancer.

Non-stick cookware when heated, even at relatively low temperatures, releases Perfluorooctanoic acid (PFOA), linked to thyroid disease, infertility, and developmental and reproductive problems. The Madrid Statement, that was signed by more than 200 scientists from 40 countries, presented the scientific consensus on the harm of PFAS chemicals.

I don't cook using plastic or non-stick anything anymore. These studies have made me a lot wiser and far more cautious about what I cook and ingest.

Cotton buds/swabs

- These are horrific polluters worldwide as they are often put down toilets and end up on beaches.

- Most are now made with paper or wooden handles here in the UK but often come in a plastic box.
- They now come in cardboard in plastic free shops and online businesses. They care!
- LastSwab™ is a forever one made of Silicone, kept in a small handy non-plastic box, that is washed after use. (Better than disposables.) They do one for Babies now too. Make sure you do get the original as they have stated on their website that they are being copied and that the quality is therefore poor.
- You can also get an Ear-spoon/Ear Loop, that has been used for thousands of years by many ancient cultures, to clean ears. Always follow the instructions for safe use.

Crayons

- Wax crayons are made of Paraffin wax, derived from Petroleum, with added colour pigments.
- Paraffin wax is a by-product of the oil purification process. So, crayons are made of a fossil fuel.
- Beeswax crayons are available. They are non-toxic, safe for all artists young and old and the environment.

Crisps

As we love them so much here in the UK and there is a mountain of plastic rubbish created by them: **Twofarmers.co.uk** of Herefordshire make their 100%

plastic free crisp-packets from plant material that they claim will break down in home composting units in just 26 weeks. Their delicious crisps, made from their own potatoes grown on the farm, come in bulk-boxes and sharing tins.

They source local ingredients for their crisp-flavours, which have no:

- Artificial flavourings
- Artificial colours
- Preservatives
- And are gluten free
- Their Lightly Salted, and Salt and Cider Vinegar flavours are suitable for Vegans too.

Cutlery - See Picnics and food on the go

Curtains

So many are plastic with a plastic backing.

Alternatives to the polyester or mixed fibres are:

- Cotton
- Wool
- Flax
- Hemp
- Silk
- Some companies do a bespoke/handmade to measure service.

D

Deodorants/Antiperspirants

We can use:

- A Natural Alum Stone – have to wet this.
- Minerals in tins.
- Natural creams in glass jars and tins.
- Natural plant-derived powders with essential oils, in cardboard compostable tubes.
- Lavender essential oil can be used in an emergency as it is antibacterial, smells fresh and is gentle on delicate armpits.

Deodoriser

- Do it the natural way with Apple Cider Vinegar. It is known to be antibacterial and can eliminate bad smells.
- When mixed with water and Epsom Salts to make a foot soak it can help get rid of unwanted foot odour by killing off the bacteria.
- Mix Apple Cider Vinegar with water in a glass spray bottle to make an air freshener.

Use Apple Cider Vinegar the way you would any White Vinegar for cleaning. It has a much more pleasant smell that clears fast, more apples than vinegar. It is affordable and can be sourced in the UK in a glass bottle but it may still have a plastic pourer and plastic inside the lid.

(Sometimes we just can't win!) It has many uses around the home and there is a lot of information online.

Dishwasher, homemade, cleaner

Avoid the nasty chemicals in plastic with this recipe:

- 1 cup of White Vinegar. (Source in a glass bottle if possible.)
- Place the cup of Vinegar on the top rack of your empty dishwasher.
- Run a hot wash.

1 cup of Bicarbonate of Soda can be used instead:

- Sprinkle across the bottom of the dishwasher and run a hot wash.
- You can add common garden Thyme to White Vinegar and let it infuse in a jar or bottle for a few weeks before you use it for cleaning. This herb has germicidal and antibacterial properties. It can be used to clean kitchen and bathroom surfaces. It can also be added to your laundry in the rinse cycle, and used in dishwashers.

Planet-friendly dishwasher tablets/tabs

These are natural and biodegradable, so safer for the environment. Some have an inner film coating that will dissolve in the dishwasher. Most eco-friendly pods/powders come in a cardboard packet and some are Vegan as well.

Dish cloths don't have to be made of plastic fibres.

- Old cotton t-shirts make good free dish cloths.
- Cut any old soft cotton fabric to size and keep it in a Rag bag in the kitchen or somewhere else dry and convenient.
- You can knit or crochet a dish cloth that is a little more robust from heavy cotton string that is environmentally friendly and biodegradable.

Dog-poop bags

These are the worst invention ever, in my humble opinion! The 'stick and flick' method worked for decades and now we have poopy bags left around, on our streets, or 'decorating' trees, by idle people. A huge problem was created almost overnight by this UK law, placing an unfair financial burden on dog-owners as well. There are inexpensive bags available made of corn starch which are compostable and 100% biodegradable that can be put in the dog-poop bins. So, there is no need to use a petrochemical plastic bag for 'poop-harvesting' any more.

Dry cleaners

A great deal of plastic is used in these businesses but some are going down the more Earth-friendly route now, and also using natural chemicals instead of the highly toxic cleaners. These are the sort of concerns we should be supporting.

Urbag is a game-changer in the industry as it is a reusable/ returnable garment bag/tote bag made of recycled plastic bottles, that would have gone to landfill, and cotton. It's dropped off as a laundry-bag then cleverly converts to a suit-bag for the return of the garment.

Readers know by now how I feel about recycled plastic but this is definitely better than the alternative scenario of tonnes of thin plastic garment-bags being used worldwide.

Dusters

Try to avoid:

- Nasty plastic wipes
- Chemicals
- Microfibre cloths
- Long-handled fibre dusters that shed microplastics everywhere.

Alternatives:

- Keep a Rag bag of old soft cotton for dusting.
- Steam and a cotton flannel, or just a hot flannel, removes grease from surfaces. Wash the flannel with hot water and bar soap and leave to dry, or throw it in your regular hot wash.
- Feather dusters are still available and do a grand job around delicate items. They are usually a by-product of the chicken and duck-meat industries so might not be an option for Vegans or Vegetarians. You can find cruelty-free feather dusters, as well as cushions, pillows and duvets, etc. Feathers are gathered during natural moulting cycles. All birds go through these vital periods of renewal annually to shed old and damaged feathers.

E

Easter eggs/Bunnies/Chicks

Most Easter treats come in thick plastic clamshells. Although these might be recyclable, we must try to avoid anything encased/boxed/wrapped in plastic when we can. Some UK producers have ditched the plastic and now favour the following materials:

- Aluminium-foil-wrapped chocolate.
- Real Cellophane wrap made of plant-material that can be composted.
- Recycled card boxes.
- Recyclable card boxes.
- Home compostable or recyclable materials – used by Eco-chocolatiers.

With chocolate that is Fairtrade – from farms that treat workers well and give them a living wage.

- Traidcraft stock Fairtrade chocolate too.
- You can find plant-based, Vegan chocolate/eggs/bunnies/chicks in many zero waste/plastic free businesses.

Eating when out and about out

- Try to avoid eateries that use Styrofoam food-containers that are no good for us or the planet!
- Ask for no straw when placing your cold drink order if all they use is plastic.
- Take a plastic-free cup/mug for hot drinks and a tall stainless-steel mug for cold ones just in case

they can be used. Some eateries will refuse to use a customer's own.

- Always carry a straw if you need one. There are different types available now. (See **Straws** in the **Swaps** section.)
- Take your own cutlery, and chopsticks, if required.
- Take your own cloth napkin that can be washed to avoid the paper ones.
- Take a 'doggie bag'/container for any leftovers to avoid waste.

Eggs

- We should try to avoid eggs in plastic.
- Buy free range eggs in cardboard if possible.
- The cheaper the eggs in plastic, the less likely they are to have been humanely farmed.
- Also make sure the hens have been vaccinated against Salmonella. (I was once terribly sick from a bad egg!) There will be a label or wording on or inside the box to this effect.

Eggshells

- Wash thoroughly with boiling water to sterilise them so there are no contaminants.
- They can be spread out on a tray in a just-used oven.
- Crush and place onto the soil around plants or in planters to deter slugs and snails. They don't like anything that will scratch their delicate under-parts.

- Eggshells add Calcium to the soil.
- Some birds will eat them in Spring to give them a boost of Calcium for egg-laying and generally as an aid to digestion.
- I think they make an attractive addition to the top of plant-pots instead of using gravel which has to be dredged from our river-beds or pebbles that come from sea-beds. Neither is considered sustainable any more.

F

Fabric softeners/conditioners

These are bad for clothes, our health, and the planet. There is no reason to use these in liquid or dryer-sheet form as they contain the following nasty cocktail of ingredients.

- Quaternary ammonium compounds which are linked to respiratory and skin irritations and are also harmful to marine environments.
- Petroleum derived ingredients.
- Plastic microcapsules full of scent that end up in our waterways and oceans.
- Phthalates – endocrine disruptors.
- Preservatives – linked to skin-irritations and cancer.
- Artificial colours.
- Palm oil that is causing widescale deforestation and the destruction of whole ecosystems in some parts of the globe.
- Tallow – in some, which is animal fat. Vegans beware!

Fabric softeners aren't good for your washing machine either and contribute to a build-up known as 'Scrud', which can then be deposited onto your washing. They also reduce the moisture absorbency of fabrics, which is a problem for any article made of towelling. Laundry that has been softened/conditioned is more flammable and may catch fire if caution is not used.

There are natural and safe alternatives such as putting in your regular washing powder and then adding a quarter to a cup of Bicarbonate of Soda to the drum, depending on the size, as it softens the water and therefore the clothes, and deodorises. Bicarbonate of Soda can be sourced in cardboard in the UK from some supermarkets and online.

You could also use Wool dryer balls in the dryer that:

- Reduce static and wrinkles.
- Restore laundry naturally.
- Don't have any harmful chemicals or scents in them.
- Are long lasting.

Face and hand wiping

This shouldn't mean reaching for the plastic wipes every time. There are natural wipes available but most still come in a plastic packet or box at this time.

Alternatives:

- A soft cotton flannel that can be washed.
- Soft cotton cloths you make yourself that can be washed.

- Old t-shirt rags are great for a quick wipe when out and about and can be thrown away or washed as necessary.
- Soft damp cotton squares or a flannel kept in a wet-bag or a reusable long-life plastic tub with a secure lid. (This is how my mum coped with her dirty little monkeys!)
- Use what you already have when you can.

Face-masks/face coverings

I didn't think I would ever be writing about these!

It is estimated that a staggering 4.2 billion single use masks are used every day throughout the world at this time. Check on the latest World Health Organisation advice to see which type of mask you should use and follow your government's advice too. It is so confusing for everyone, as advice from all quarters keeps changing as the pandemic evolves.

If you see discarded single-use gloves and masks and are tempted to pick them up, use a long-handled litterpicker and put them in the nearest bin or straight into a bin bag. They endanger wildlife, especially the strings on masks that entangle birds and sea life. These strings should be cut before disposal. I know a lot of Eco Warriors have been litterpicking and cutting strings to help wildlife in their local communities, and I want to thank every one of them for caring!

Polypropylene hospital medical-masks can be recycled back into plastic building blocks, which absolutely horrifies me. What used to be incinerated in the hospitals, and gone, is now more plastic polluting the planet!

Alternatives to single use masks if you don't require medical ones:

- Commercial soft all cotton/washable, with the WHO recommended 3-layers.
- Handmade cotton/washable with cotton loops or around the head ties – lots of instructions online.

Buy or make a few face-masks so you always have a clean one to hand or keep in the car. You might like to keep them in a dedicated zip-pouch or purse somewhere handy too.

As we all need workable options, Last Mask from the company Last Object offers a neat little dispenser pack for one 3-layer cotton mask incorporated into a refillable Silicone spray-dispenser for your preferred sanitiser.

Leave a note by the door:

'Got your mask?'

On returning home:

- Remove mask by the loops.
- Wash hands first and then the reusable mask in hot soapy water, rinse thoroughly and leave to dry. You could put it in the washing machine. (I

personally prefer the former way as it is over and done with fast and the mask won't get forgotten for the next time. I keep 5 on the go and wash them in a dedicated metal mixing-bowl.)

- Thoroughly wash hands again afterwards using hot water and soap. It will become routine.

Be safe!

Felt-tips/highlighters

If you're trying to avoid plastic felt-tips, buy some lovely watercolour pencils instead. They can be used as pencils and are also great fun as paints. No plastic or recycling involved! They are better for health and the environment and can be purchased from art suppliers. Jumbo watercolour pencils are becoming popular for little hands. Just make sure you get a good metal sharpener.

Feminine sprays

We women don't need them. It's just another way the chemical industry relieves us of our hard-earned cash and makes us feel bad about our bodies.

Female bodies function beautifully without adding anything that could disrupt the natural balance. Just keep clean using a cotton flannel and an unperfumed soap, as perfumed ones can lead to irritation. If there is anything unusual going on 'down there', seek medical advice.

Fitness/Yoga mats

These are usually made of plastic, especially if they are Latex free, but there are natural alternatives that are becoming more readily available.

Look for mats made of:

- Cork from the Cork Oak - non-slip, soft, warm and sustainable. Available from Corkspace.co.uk.
- A combination of Jute plant fibre and real Latex rubber.
- Natural rubber from the sustainable Rubber tree.

Floor-cleaning

- If you have a steam-cleaner then continue to use that.
- If you need a new standard type wringable floor cleaner then get a good one that will last years of tough work.
- You might even be lucky enough to find a mop that has a replaceable head.
- Cotton string mops with wooden handles in metal buckets are still available, from hardware stores and online, for tile and lino cleaning.
- Wooden floors can be wiped over with a warm barely damp cloth.
- (I do mine with a just turned off steamer.)
- One enterprising person has developed a free online knitting pattern for making floor-wipes to attach to the usual flat-type floor cleaners. They can be created plastic free using cotton thread.

- You can use any old towelling tied onto the old flat mop-head. This can eventually be composted if no chemicals are used.

Floral foam for flower arranging

There was life before plastic green floral foam appeared. There are many alternative methods of flower arranging without using any, that can be found online, like projects from the London Flower School.

A non-plastic biofoam is becoming popular with flower-arrangers. Ask at your local florist if they stock/use it. If they don't then you've planted a seed in their mind about sustainability and being plastic free. You may have to source it online but even that is getting easier now.

Floss/dental floss

Most is plastic but a lot of people are rightly concerned about putting any sort of plastic in their mouths now.

These types of floss can be found in carboard or in a metal refillable container/dispenser:

- Silk – biodegradable but please check how it is produced. Most silk producers kill the developing moths in the cocoons by boiling and adding chemicals. There are Moth-safe silks now where the Silk moth cocoons are unravelled harmlessly, enabling the moths to live out their natural lives.

- Plant-based Vegan floss.
- Corn starch – biodegradable but may not be home compostable. Better to cut it up after use, in case it finds its way into the environment to harm wildlife, and then bin it.

Floss should never go down the loo whatever it is made of!

Food waste

Never put food-waste or food-contaminated boxes, etc., in domestic bins or bin-bags!

Number one rule for disposal or recycling of food-packaging: If it's dirty, wash it out first.

Food waste, even a tin that still has some food residue in it, will attract insects, animals like Foxes and Badgers, and birds of the Sea gull variety who will bite and peck holes in bin bags to get the 'food' out.

They are just hungry, poor things!

It's dirty people who are ruining our world, not the wildlife!

Organic/Food waste

Use up veggies and peels as stock and batch cook meals/ soups and freeze to avoid waste in the first place.

If that is not possible then we can:

- Compost it.
- Put it in a Worm bin.

- Put in a kitchen Bokashi bin.
- Or put it in a Council/Municipality food-collection bin.
- Remove any plastic labels beforehand.

If we keep our kitchen bin food-scrap free, it won't smell and there will be no need for any sort of lining because what is in it will be dry.

Footwear

The global industry is big business with 20 billion pairs of shoes manufactured each year. In 2020 it was worth over $370 billion, with the **Trainers** industry alone worth $58 billion in 2018, expected to rise to $88 billion by 2024.

Trainers/Plimsolls/Sneakers

England's famous monarch Henry VIII had some shoes specially designed in lightweight leather in the Sixteenth Century so he could enjoy his game of tennis, and the tennis-shoe/trainer was born. In 1849 Charles Goodyear patented Vulcanized Rubber where the natural rubber is mixed with Sulphur to make it flexible but more hard-wearing. This was perfect for the soles of shoes and later for Wellington boots/Wellies, named after The Duke of Wellington. Plimsolls, the type of shoe all UK children wore for PE when I was a child, were made of this rubber, with an upper of cotton canvas. American sneakers were made in a similar but more stylish and colourful way.

Among the treasures I brought back from my stay in the USA in 1976 were a dinosaur footprint, yes, really, and a lovely red pair of American-made sneakers, both of which I still enjoy. I wore these wonderfully comfortable shoes for the dig and for a while afterwards when I got home, and then I put them away because I didn't want to wear them out. I'm so pleased to see that canvas and natural rubber is being used once again by some manufacturers for modern trainers and sneakers.

A typical modern, highly-designed, trainer has over 65 components, a complicated mix of natural and synthetic elements that don't allow for recycling. The global waste/pollution is mind-blowing for this ubiquitous form of footwear that is produced using chemicals that have a detrimental impact on the environment during manufacture, use and ultimate disposal. We must also avoid buying any fancy light up ones as they will end in landfill or incineration. They can't even be recycled because they contain embedded electronic components.

Health experts recommend buying new trainers to the following timetable. It honestly makes me feel ill to think of the horrendous global pollution it must create if we stick to these ridiculous guidelines.

The following is how long trainers usually last before they need replacing:

- Regular walker – 3 – 5 months
- Jogger – 3 months
- Gym user – 6 - 12 months
- Serious runner – every 300 – 500 miles

I find this absolutely shocking, coming from a generation that was able to get things repaired to give them many more years of life!

It's not easy going plastic free when it comes to footwear unless you wear leather, even then the sole will probably be comprised of polyurethane whose only satisfactory option for disposal is incineration for energy, according to the industry. Some companies are including recycled ocean plastic in the uppers but to me that is the usual delay in finding a natural solution to petrochemical plastic.

Flip-flops

This type of footwear is a huge ocean polluter as millions of people wear them in the hotter climes and so many are lost in the water.

Look for water-shoes that:

- Stay on your feet if you intend visiting a beach.
- Are made of natural materials like canvas and rubber, even algae now, that won't harm the ocean when lost.

Many eco-conscious shoe brands are moving away from plastic and animal-leathers and are now looking to fruit and plant leathers as replacements.

These include the following:

- Pinatex – made from pineapple waste, with more income for the farmers.
- Bioplastic made from apple waste – again no waste and a second income for farmers.
- Fleather - made from Indian temple flowers which would otherwise go to waste. Excellent, as all worshippers take flowers to Hindu temples.

Mexican entrepreneurs Adrián López Velarde and Marte Cázarez recently debuted Desserto, the first organic leather made entirely from the Nopal, Prickly-pear, cactus that they hope will replace leather in the clothing and shoe industries.

Other natural alternatives for footwear include:

- Shoes made from algae or eucalyptus pulp.
- Slippers made from wool, with a natural rubber sole – how they used to be made before plastic.
- Sandals made from cotton canvas, with a jute, plant fibre, sole.
- Espadrille or Espardenyes - casual shoes made of a canvas or cotton fabric upper and a flexible sole made of Esparto rope. This rope sole is the defining characteristic of an Espadrille but the uppers vary widely in style and shoes can be flat-soled or wedge-shaped.

Esparto - Halfah grass or Esparto grass is a fibre produced from two species of perennial grass from North Africa and Southern Europe. It's used for crafts such as basketry as well as the iconic rope Espadrille.

Whatever our footwear is made from, we need to return to repairing it if we want to reduce waste. Shoes are often poorly made and only last months instead of years now and are binned when they are finished with or out of fashion. We must buy to last to avoid waste.

I've gone through a dozen pairs of shoes in as many years because of the poor build quality. I don't drive so I walk a lot. When I was young, we wasted nothing in a still partly devastated UK. Clothes were sewn and knitted by most mums and grannies, and other things were mended and repurposed by many dads and grandpas. Those included shoes, as it was easy to buy soles and heels and the specialist glue needed to fix them on. I remember my Father had a heavy Cobbler's last for this purpose and often disappeared 'down the shed' to mend the family shoes. Later on, when he became too busy and we had a bit more money coming in after my mum started work again, we used to take them to the Cobblers in town.

Now the buzzword is 'circular-economy' but that is nothing new. Years ago, when our shoes were well made, they could be constantly soled, heeled and generally tidied up in these specialist shops, the Cobblers, which were found in most towns. They also repaired bags, luggage and belts, that are often just thrown away now if spoilt, damaged or out of fashion. Shoe-making and mending are ancient skills that sadly have almost been lost to 'progress' and the onslaught of plastic. I remember taking a family bag of shoes down to our local Cobblers for an overhaul. They always came back well repaired, highly polished and looking like new. It was a relief to know that favourite comfortable shoes could have a second and even third life.

We need well-made, plastic-free shoes and the Cobblers to care for them!

Freezer/Food bags

Home compostable bioplastic bags are available from plastic free/zero waste businesses.

Fruit and veg in plastic

Loose produce tends to be bigger in the supermarkets and is often of a higher quality. It's best, though, to buy from farmers' markets, farm shops and greengrocers if you are able to support your local economy.

Buying in season is what we should aim for.

Fruit nets

- Avoid buying fruit in plastic netting whenever possible.
- Cotton netting is different because it can be composted at home.
- We should never use plastic netting as a pot-scrubber as microplastic goes down the plughole and out into the waterways. It is then consumed by water-life and will eventually damage/poison or kill it.
- Better to bin plastic netting and then avoid buying any fruit in it next time.
- Look out for loose fruit or those in cardboard boxes.
- Some UK supermarket-chains are, thankfully, trying to persuade their producers to find more natural ways of packing their fruit but it will take time to become the norm.

Furniture polish/spray when dusting

- Avoid the wipes and sprays.
- Don't use plastic-fibre dusters that shed particles constantly.

- A slightly damp old cotton t-shirt rag – free and chemical free – will do a good job and can be rinsed out, dried, and used for years.
- Beeswax that comes in a tin, applied sparingly, is a good substitute for any of the sprays if you really need to use something. It also smells divine and gives a wonderful shine. It was used for centuries before chemicals appeared. You will also be supporting beekeepers.

G

Gardens and gardening

We must reconnect with the soil and its riches and eliminate all plastic from the process. I'm a passionate natural wildlife gardener and know how difficult it is to do this but we can make a start in our own gardens, and then pester suppliers and producers to do more. We can try to source plants in natural containers and only use terracotta/ceramic or metal/aluminium pots, ceramic sinks, wooden planters and barrels in our own gardens. I love upcycled natural materials. They make quirky and attractive features for any garden.

Coir is the hairy part of the coconut that is usually discarded that can be used to make good plastic free alternatives for many things in our gardens such as:

- Pots
- Weed control mats
- Planting pods
- Growbags

- Nursery bags
- Hanging basket liners that are better for the plants, drain well but still retain moisture.
- Boot scrapers
- Outdoor and indoor mats

Hanging basket liners can also be made out of:

- Hessian from old sacks – made of Jute plants
- Old cotton flannelette sheets and pillow-cases
- Old wool blankets
- Old cotton tee-shirts
- Wood-chips

Green or beige recyclable plastic pots are being developed to cut out the use of black plastic pots that can't be recognised by the scanners in the recycling depots, therefore ending in landfill or incineration. Not ideal but there are trillions of black plastic pots on the planet. More must be done by all the gardening/growing industries to address this Everest of a problem, that we gardeners are acutely aware of and can do little about at present.

Try to source any garden products such as top soil and compost/manure or mulch, etc., plastic free if possible. Remind any business you have contact with that single use plastic, even if it is thick and can be reused, is a no-go area of gardening now. The world can't take any more plastic and gardeners must adapt to life without it, again.

We must avoid buying Peat and look for Peat free compost.

A Peat bog is a rare ecosystem that purifies any water flowing through it, and helps to prevent flooding. Harvesting Peat destroys this delicate, slow-to-regenerate, ecosystem that has developed over aeons and consists of decomposing plant material.

Recycling old tyres as planters has become popular but they are still in the environment doing their nasty toxic thing, especially when the weather heats up. The 'attractive' painted ones are even worse as the paint is probably toxic too. Tyres are one of the most common plastic polluters on the planet. They are often found dumped by the thousands in oceans. I therefore wouldn't advise using them if you are trying to go plastic free in your outside space or garden.

One more thing that worries me is the plastic plant labels/tags left on new plants and trees, that deteriorate over time into plastic fragments and, worse, microplastics. It is therefore a good idea to remove any straightaway and store them somewhere safe so you still know what you planted. We gardeners often completely forget what we have planted!

Old pallets make good homemade compost bins and planters if you have the space. They can also be used for fencing and outdoor and indoor furniture and shelving. Plenty of ideas online and in specialist books.

Greenhouse glass cleaner

Use:

- Regular White Vinegar or Apple Cider Vinegar in a glass spray bottle, and rinse with warm water.

Reuse an old plastic sprayer if that is what is available.

- Scrunched up newspaper has been used to clean glass for years and is most effective.

More in **Planters/plant pots for the garden and house**

Gift wrapping

We can use the following alternatives to the usual plastic 'paper' and 'ribbon':

- Brown paper from parcel-packaging, decorated with dried natural foliage and pine cones, tied with Jute string is very attractive.
- Raffia, that derives from the leaf-veins of the Raffia Palm tree grown in tropical regions of Africa, especially Madagascar, and Central and South America, is very pretty. It makes a lovely alternative to ribbon and comes in different colours and widths.
- Old sheets of music
- Old magazines
- Old comics
- Maps
- Pretty seasonal tea towels will make a bonus gift.
- Old cut up sheets.
- Use up all those old newspapers hanging around.
- Pretty cotton pillowcases or small tablecloths – charity/thrift shops often have a good selection of both.
- Any old cotton clothing cut into manageable pieces – soft cotton is easier to tie into bows for closures.

Christmas homemade drawstring gift bags instead of wrapping paper:

- Large ones: 50cm x 50cm
- Second hand material is obviously more eco-friendly if you can find it. Even better if the pattern is a bit Christmassy.
- Can be used for many years in the family.

Santa sacks
Use a pillowcase. That was all we had until the American idea of specialised sacks came over to the UK.

Fabric wraps instead of paper wrapping
Furoshiki style Japanese wrapping is great for gift-wrapping at any time, especially for bottles and boxes. The result is a wonderful-looking present and a bonus gift of the reusable fabric or scarf. Lovely head-squares/scarves can be found in charity/thrift shops.

There are many videos for all these ideas online, with ingenious solutions popping up daily.

Giveaways/plastic toys

Just refuse them, and don't buy magazines that have them on the front. Most end up as landfill or in our waterways and ultimately oceans.

Gloves

A lot of gloves are now made of some form of plastic fibre or faux leather that is plastic.

You can look for these alternatives if you don't want to go for leather/sheepskin:

- Thick organic cotton
- Wool
- Cashmere wool - from the Kashmir goat

Turtle-doves.co.uk collects Cashmere clothing, turning it into fingerless gloves/wrist-warmers, that are especially good for Arthritis and Raynaud's sufferers, scarves and shawls. You can even send them your old Cashmere sweaters to help with recycling this precious natural material.

Gloves at petrol stations

These are unnecessary and are adding to our plastic nightmare. Often mindlessly discarded on forecourts, they find their way into watercourses on a windy day. Please don't use them. Motorists were fine years ago when these weren't even a money-making dream. Just wash/cleanse hands as soon as possible if needed or keep some natural cleanser in an aluminium pump-bottle in your vehicle. Beautykitchen.co.uk do a lovely one that is not a gel but a gentle mist.

Gloves for washing up and household use

You can still buy reusable latex rubber gloves, as long as you are not allergic to latex, called **'If you care'**. They are available from many eco-businesses and are completely planet-friendly, Fairtrade, Certified Natural Rubber, and should break down naturally in a composter or if they accidentally enter the environment.

Latex household rubber gloves are also reappearing in many supermarket chains that are trying to reduce their stock of plastic products.

Golf tees

These awful plastic polluters don't need to be everywhere. They can be sourced made of wood or Bamboo that will naturally rot down if they are lost in the environment.

Greetings cards

Try to avoid any musical/light up ones that can't be recycled because of the electronics inside.

Look for:

- Regular cards with no glitter or embellishments or in plastic.
- Cards made from recycled paper.
- Cards made with natural materials, including the packaging.
- Paper cards. These don't need to be boring. There are a lot of lovely crafted ones out there if you don't make your own, and plenty of ideas online and in specialised craft books if you do.

H

Hairbrushes

All types and sizes usually have plastic bristles and a plastic cushion.

Look for Bamboo-handled brushes with Bamboo or other plant-based bristles, available from plastic free businesses. They are becoming popular now and demand is soaring. I use a lovely Bamboo one with a rubber cushion and Bamboo bristles. It's pleasant to handle and does a really good job, and was great value for money for a lifetime purchase.

Hair dye

Millions of people use some form of hair-dye daily. Chemical hair-dye is one of the world's biggest river and ocean polluters. If you love Earth, avoid the dye! The chemicals in dyes and all the plastic paraphernalia that accompanies hair-dyeing contribute to killing entire river-systems.

We should be using natural dyes, like real Henna, from a plant, that have no harsh chemicals in them that harm us or the environment. You can also get long lasting natural hair dyes that are really easy to use as you simply add water to the reusable applicator bottle, shake and apply.

Scientists in the UK have even successfully created red, purple and blue natural long-lasting hair dyes from Blackcurrant waste left over from food and juice processing. When a natural yellow colouring is added, they can also create a range of brown tones.

Natural organic hair dyes are free from:

- Heavy Metals
- Ammonia
- Peroxide

- PPD – p - Phenylenediamine, also used in Kevlar body-protection products

Beware of fake Henna products!
'You're particularly at risk if you have (or have previously had) a black henna tattoo.' (Real Henna paste is orange/red and not black. Black Henna is not true Henna and therefore not natural.)

'These temporary tattoos should be avoided because the paste often contains high levels of PPD, which can increase the risk of an allergic reaction the next time you're exposed to it. So, you could develop a life-threatening allergic reaction when you next use PPD hair dye.' – NHS, UK

Better to be safe than sorry and avoid any hair dye with PPD in it. Always check the packaging.

You could of course give it up altogether and embrace the grey. Better for you, as it will save you a fortune, the palaver and mess, and for the planet. No more chemicals going down the plughole either. You can then spend the money saved on more plastic-free essentials and good plastic-free food.

Hairsprays

Although they come in aluminium cans, they usually have a plastic lid and sprayer and are full of chemicals.

These include among others:

- Plasticizers
- Alcohol that is drying.

- Lustre-agents
- Gas-propellants that are no good for us, our pets, or the environment.

There are recipes for natural alternatives online that you can make at home if you can't source any plastic free. Look for the Leaping Bunny Cruelty-free international logo if buying a commercial one.

Hair ties and bobbles

The majority of these are still made of plastic but you can get organic cotton and real rubber ones. Tabitha Eve, who make 'NObbles' – NO(plastic-hair-bo)bbles, in the UK, and Kooshoo in the USA and Canada specialise in these.

Halloween/Trick or Treat gifts

Avoid all the usual nasty cheap plastic in gift bags with these alternatives:

- Homemade spooky-shaped biscuits/cookies – have fun!
- Pencils and crayons – give them a spooky twist!
- Activity sheets – be creative with fun pictures to colour in.
- Make some real cotton and wire pipe-cleaner spiders!
- You can create little creatures such as owls out of pine cones.
- Chocolate coins, or eyeballs, in foil.
- Fresh seasonal fruit.

- Toffee/Caramel apples.
- Nuts.
- Decorate the house with natural things that you could find on a walk in the woods. There are lots of ideas online and in specialist books on having a natural Halloween.
- Avoid the spray for spiderwebs, it's plastic.
- It's easy to make spiderwebs with string, instructions online. Put cotton pipe-cleaner spiders in the middle. Spiders come in all sorts of colours so don't have to be just black. Have spooky fun creating lots of different ones!

Pumpkins should be eaten and not just used for decoration.

Seeds can be roasted, instructions and recipes online, and the delicious 'Turk's Turban' squash makes lovely muffins, soups and vegetable stews. Any pumpkin parts left over can be put in the garden for little creatures to enjoy too. Squirrels love them! Put them up high out of reach of our precious Hedgehogs, though, because they are not good for their little tummies.

Hand creams/body lotions and potions

We should try to avoid the ones that contain petroleum jelly and palm oil and look for natural ingredients such as:

- Cocoa butter
- Shea butter
- Jojoba oil
- Orange oil

- Calendula
- Evening Primrose oil
- Chamomile
- Grape seed oil
- Sweet almond oil
- Peach kernel
- Lavender
- Geranium
- Cedarwood
- Rose oil
- Hemp balm

Hand sanitiser

According to scientific research, most is hormone-damaging gel in plastic. Just remember to wash hands thoroughly with hot water and soap when you get home if you are unable to wash them when out and about or at the first opportunity. Hand-cleansers/sanitisers that contain only natural ingredients are available from eco businesses, shops and online stores.

I use the refillable one from Beautykitchen.co.uk. It has a pleasant smell and is a spray, not a sticky gel.

Handwash in a pump

Some people just prefer using liquid soap. They can often be converted to soap bars in the bathroom but prefer a liquid pump-dispenser in the kitchen. The vast majority of commercial liquid hand soaps are in fact detergents and come packaged in small plastic pump-bottles that we're trying to avoid. Natural soaps in

Aluminium pump-bottles are available. You can now also get liquid soaps from dispensers in some refill shops. Just take an old bottle. Sometimes you can buy one there.

Bar soap is having a much-needed revival and is becoming popular once again. Lots of varieties are available with less perfume or perfume free for those with sensitive skin, and for the gentlemen in your life who don't particularly want to smell like flowers! Personally, I think we could do with a few more fragrant men!

We only had bar soap when I was a child and it did a great job cleaning us and all around the house. We even sang a silly innocuous little ditty about it:

'While shepherds washed their socks by night
All seated round a tub
A bar of Sunlight Soap came down
And they began to scrub!'

Advice, gratefully received, from Emma's Earth-Conscious Living on Facebook, using safe dilution guidelines for essential oils as stipulated by the cosmetic safety assessors.

A simple, two-step, Sweet Orange Liquid Soap recipe using Organic liquid Castile soap, which is about as gentle and pure as you can get when it comes to a natural cleaning agent.

It is:

- Vegetable derived.
- Fragrance free.

- Suitable for use on hands, body, face and hair.
- Used to wash dishes and can even be used on wooden floors.
- Often stocked in zero waste stores for you to refill your existing containers, or can be purchased in bulk (5 litre bottles) - a huge reduction in single use plastic.
- Unscented so you can add Essential oils for their gorgeous fragrance and skin-kind properties. As always, these must be added in safe quantities, at a maximum of 3% concentration.

(Emma has found that 2% is ample). Sweet Orange is her Essential oil of choice - a relatively inexpensive oil, suitable for use on children and with an unmistakably fresh and zesty fragrance.

To make one bottle of this natural liquid soap:

- Add 5ml of Sweet Orange essential oil to 250ml liquid Castile soap.
- Mix well. The soap will cloud and colour slightly due to the addition of the oil - this is normal.
- Pour into a liquid soap dispenser of your choice, avoiding plastic containers where possible, as this preserves the integrity of the Essential oil. Glass or stainless-steel is best.

It's an Earth-Conscious liquid soap that can be made in minutes and is suitable for use on hands as well as in the bath or shower. It is nice to know that your soap is kind to the planet and kind to your skin.

Felted soaps
These, made with a wool covering that is a built-in exfoliator, are also becoming popular now. They can be purchased from specialist makers and there are instructions online for making them yourself. Some plastic free and zero waste shops also stock them.

Heated tongs and straighteners

Most tongs have plastic handles or contain some plastic. All usually end in landfill even though electrical personal grooming appliances should go in the electrical waste at the tip for recycling. It's therefore best to buy good appliances that will last. If an appliance has a plug, battery or cable it can be recycled in the UK.

Before the iron wand heated on the stove and later modern electric hair styling tools, women used strips of cotton rag to curl their hair or created pin-curls with hairclips. Doing this once in a while would give hair a rest from heat-damage and be an interesting trip back into the past for a vintage look. Instructions for these techniques can be found online.

Honey and Jams

Go for glass jars to cut out the plastic. Some have a plastic anti-tamper label or collar on them but that is better than all plastic.

Hot water bottles

You can still buy real rubber hot water bottles, and they last decades. I have one. Just look for the label stamped onto the neck. Sadly, the screw top will probably be made of Nylon plastic. It was metal years ago, with a natural rubber washer.

I

Ice cream when out and about

- Choose a cone or wafer to avoid the plastic cup and spoon.
- Eat in to use the glass or ceramic dish supplied.
- Avoid all the different kinds of lollies/popsicles in plastic too. They create piles of plastic-wrapper rubbish all over the planet!

You can still get lovely stainless-steel ice cream/lolly/popsicle moulds for DIY ices at home using fruit juices/cordials.

Insoles for shoes

- These are usually made of a mixture of plastic materials, and are littered all over the globe.
- An all-natural alternative is Cork – from the sustainable Cork Oak – and was used for insoles before plastic ones. It's naturally antibacterial and comfortable in footwear. It can easily be cut to size too.
- Cork insoles are becoming popular once more, after almost disappearing from the shelves, but you may need to get them online. Years ago, every high street Cobbler, shoe-mender, and supermarket stocked them.

Inflatable pools, paddling pools and beach toys

Wyatt and Jack (on Instagram) collect them and make them into high quality tote bags. What else can we do with the millions of inflatables that end in landfill or, worse, rivers and oceans other than use them in better-than-landfill projects like these if they continue to be manufactured?

We must therefore try to avoid buying them for the sake of the planet. Naturally, I would like to see non-petrochemical plastic alternatives made of biodegradable materials but that is a long way off at present. Years ago, they were made of biodegradable rubber. Maybe we could go back to this in a more modern form.

J

Jewellery

- Fashion jewellery is usually synthetic/plastic. Avoid it when you can.
- Go for natural materials that will last many lifetimes.
- Ceramic, and glass jewellery can be very attractive.
- Semi-precious stone/crystal jewellery is gorgeous and affordable as well as having health benefits and has been used for thousands of years by many ancient cultures. A whole slew of books has been written on the efficacy of these natural wonders. You can find beautiful crystals/crystal jewellery, often in Silver, in New Age/Spiritual shops and online businesses.

- Buy second-hand whenever possible, as all metals and minerals take huge resources to extract and process and we should be thinking planet first.

Jug kettles

- These are a combination of plastic and metal.
- Heating anything plastic is not a good idea because of the toxins it can release.
- Stainless-steel is the best option for kettles as it heats up fast and is easy to keep clean.

I use a stainless-steel pan for boiling water. My old jug kettle blew up with a loud bang! Thankfully, I had just turned away to the sink. I was shaky for hours after that and vowed no more electric kettles, of any description, for me!

Junk mail

- Millions of precious trees are cut down just for this invasive form of advertising that no-one really wants!
- A lot of plastic is used in the printing process and for mailers that are often just single-use plastic bags.
- Stop this in the UK by signing up to: **https://www.mpsonline.org.uk//consumer/what_is_mps**
- In the USA you have to ask the senders individually to stop sending mail. This is therefore a much more complicated process, but will be so worth it to save forests and general waste.

K

Kettles that are better for the environment

There is a huge variety of greener kettles now that use less water for boiling a small amount, less power, and less or no plastic.

Kettle-cleaning

- Avoid the cleaners in plastic using 1 tablespoon of Citric Acid - can be sourced in cardboard in the UK.
- Half fill the kettle with water.
- Add the Citric Acid.
- Boil the kettle and leave it standing for an hour.
- Rinse.
- Add fresh water.
- Boil again.
- Pour out the water for a nice clean fresh kettle.

Keyboard-cleaning the natural way

Did you know that keyboards can be dirtier than a public toilet, especially shared and family ones? It is therefore important that they are cleaned on a regular basis without resorting to a plastic wipe.

Turn off the computer and unplug before any cleaning.

Method:

- Put White Vinegar, or Apple Cider Vinegar if you don't like strong smells, in a small amount of

water in a cup. You can add a few drops of, strong-smelling, Tea Tree essential oil for an extra whammy of potent germ-killing if you like.
- Dab in a small soft cotton cloth/old tee and wring it out so it is damp but not wet.
- Then clean everything, including the Mouse if you use one, and not just the keyboard.

Remember: Plastic holds onto bacteria.

Kitchen utensils – alternatives to the plastic or Silicone ones most of us use every day:

- Wooden ones last for decades and have naturally antibacterial properties.
- Bamboo also lasts for decades and has antibacterial and antimicrobial properties.
- Stainless-steel will never need replacing as it lasts many lifetimes.
- Corn or Wheat plastic utensils are being considered by some catering companies.

Kites

We need to go back to flying kites that are made of natural materials. Kites can get lost in strong winds and if they are plastic, they could end up being ingested by wildlife just as balloons can. As they also have long strings attached, they could cause a similar problem to balloon-strings that entangle. Look for kites made of cotton materials with strong natural strings or make your own. Instructions can be found online or in books on kites.

It's so satisfying, and a lot of fun, building your own kite with colourful materials, collected wooden sticks or Bamboo. I used to love flying kites when I was a child camping with my family. They were cloth, the string was strong cotton and the handle/winder was made of wood and metal.

Knitting yarns

A lot of them are 100% Acrylic but you can still get real wool that has been naturally dyed or has been left totally undyed for a more rustic appearance. Often artisan makers spin their own wool for sale or for their own knitwear.

L

Laundry bins/baskets

More often than not these are plastic.

Go natural with:

- Willow
- Bamboo
- Rattan
- Seagrass - flowering plants which grow in marine environments that can be sustainably harvested.

Lemon and Lime juice

Look for these in glass bottles. Most still have a plastic lid and pourer but that is better than 100% plastic. Avoid the plastic 'Lemon' or 'Lime'.

Lint-rollers

Most are plastic with a sticky plastic surface. Some are replaceable with more sticky plastic that just has to be binned! Avoid these by using a sustainable wooden Lint Brush with Rubber Bristles. Especially useful if you wear a dark suit for work. Keep one in a drawer. They come with a pretty hefty price-tag but they will last a very long time.

Lip balms

Most are Petroleum-based and come in a plastic tube.

Look for natural ingredients, in a push-up carboard tube if available, like:

- Castor seed oil and beeswax
- Olive, sunflower, jojoba and rosehip oils

Beeswax lip balm is a popular choice now and can be purchased in some cases directly from the beekeepers. This is the best way to buy it as you're supporting a vital industry that helps secure the future of our precious Honeybees.

Liquid soap

Use bar soap in paper or cardboard. There is a huge variety available now in stores and in specialist/zero waste/plastic free businesses.

Loo roll/toilet paper

- This can now be sourced in paper or biodegradable packaging. Some is made of

Bamboo or Hemp, from fasting-growing plants and not slow-growing trees.

- Some people prefer a bidet or a bidet attachment.

Loo/toilet cleaning - A chemical-laden toilet rim block is not necessary.

- Avoid the plastic too and save money.
- We should avoid any toxic chemical cleaner in a plastic bottle.
- Natural cleansers are better for our health and the planet's.
- Citric Acid can be bought in cardboard in the UK.
- Bicarbonate of Soda can also be bought in cardboard in the UK.
- Vinegar can be sourced in glass. (If you can't stand the smell, use lemon to wipe and freshen seats, using an old cotton t-shirt rag, and then wipe again with clean water so no sticky bums/butts!)
- Apple Cider Vinegar is less smelly than White Vinegar but just as good.
- Toilet bombs, which can also be used to freshen drains, made with natural minerals and essential oils, are available in the UK from plastic free/zero waste businesses. Or you can make your own from instructions online.
- Pumice, a natural volcanic stone, used for stubborn limescale stains, can be sourced from chemists or pharmacies. You can usually find it loose, or attached to a long wooden handle, in the footcare section of Chemists.

- Natural fibre toilet-brushes with wooden handles can be found in plastic free/zero waste shops and online stores in slim aluminium buckets. Natural brushes don't hold onto bacteria the way plastic ones do if you keep them clean and dry.

Lunchboxes

As mentioned earlier, Christmas Biscuits/Cookies come in lovely tins and make great lunchboxes later. You get delicious food and a useful tin, with no huge outlay. Donate any you don't need to friends who would appreciate them or a charity/thrift shop. Win all round!

More substantial Commercial ones are available:

- Stainless-steel is expensive but will last a lifetime.
- Stainless-steel and Silicone is also pricey, but long lasting.
- Wheatgrass bioplastic looks and behaves just like plastic. I have a lunchbox made of this material for my Alternatives to Plastic Demo Table and it amazes all who see it. Wheat waste/straw is normally burned or put into landfill so this use brings new revenue to farmers.
- Corn-starch bioplastic also looks and behaves just like petrochemical plastic and is an extra revenue for farmers.
- Bamboo lunchboxes are great too but buy a good one from a reputable seller as cheaper ones might be glued with less than planet-friendly substances.

M

Make-up/cosmetic products

These, sadly, are one of the major sources of pollution.

Do your research. Read labels and look for ethical brands that use minimal plastic or don't use any plastic or plastic packaging, and are going down the refillable route. There are more joining the market daily. Some natural brands can be found in plastic free shops or online.

Look for the Leaping Bunny label to show everything is cruelty free.

Many natural products are available:

- Safe mineral foundation.
- Safe mineral blusher.
- Fruit pigmented eye-shadow.
- Eye-shadow with no glitter. Glitter is plastic.
- Mascara. Some come in Bamboo tubes but still have plastic wands that have to be disposed of properly.
- Look out for the old-fashioned 'Cake Mascara' in tins that can be refilled but check the ingredients before buying as some might not be that planet-friendly.

Method of use:

- Dampen the brush that usually comes with it.
- Mix with a drop of water on the cake

- Place the brush at the roots of lashes, and gently brush upwards.
- Allow to dry, then apply a second coat.

I remember the old 'spit and brush' method!

These Mascaras can be used as Eyeliner too.

Method of use:

- Add a couple of drops of water to the cake, and use a fine pointed brush to define the eye.
- This sort of Mascara/Eyeliner, used in Hollywood for decades, usually removes easily with warm water or a natural liquid make-up remover like Coconut oil or Jojoba oil.

Natural lip gloss can be found free of:

- Talc
- Fragrances
- Phthalates
- Sulphates
- Mineral oils
- Palm oil
- Parabens

They are available in specialist makeup suppliers and in some zero waste/plastic free shops and online businesses.

Make-up remover and pads

We used a warm flannel years ago before make-up remover was even a money-making prospect but there are a lot of alternatives now:

- Cotton rounds can be used with water or a natural liquid make-up remover like Coconut oil or Jojoba oil.
- Bamboo material rounds are becoming popular because of the worries about cotton production.
- Homemade cotton rounds or squares can be made from old soft cotton t-shirts. Squares are easier to cut and sew, with little waste. You may need to sew the middle together to hold them securely, as well as around the edges to prevent fraying.
- Always do a patch test first before using any unfamiliar product, even a natural one.

Meat and Fish

- We must try to avoid any product with plastic film around it or in a plastic tray.
- If you are lucky enough to have a local butcher and fishmonger or farmers' market, use those as the meat and fish is likely to be wrapped in paper. You can ask sellers to use your own containers.
- Remember to avoid or politely refuse any plastic bag offered if you have your own reusable one.
- Many supermarkets have meat and fish products that come in only cardboard packaging or cardboard and aluminium foil and are found in the freezer section.
- Both the carboard, if clean, and the aluminium can then be recycled.
- Wash off the cardboard and crush aluminium into a fist-sized ball so it will go through the recycling machine, and recycle the carboard with

the paper and carboard recycling. It doesn't matter if any carboard is still a bit damp as using water is part of the recycling process.

- If buying meat and fish at the deli in a supermarket, ask for no plastic film or bag and that, if possible, your own container is used.
- Some supermarkets offer containers for sale near the deli if you have forgotten yours.

Try to aim for a couple of meat, and fish, free meals per week to help the planet. The meat and fish industries are some of the world's major polluters, and meat and fish are unsustainable at this present time.

Microfibre cloths

Unless it states the microfibre is natural, it is always plastic. These cloths are damaging if they get into the environment as they release plastic microfibres. Use soft cotton cloths or rags instead.

Does you Council send waste to landfill or incinerate for fuel? If they incinerate, I would advise binning any microfibre cloths if you still have them and using more natural cloths or rags. Microfibre cloths shed forever and we breathe these fibres in each time we use them!

As usual, be safe and not sorry!

Milk – dairy

- We should avoid the plastic containers whenever we can and buy in glass.

- Use the local milkman if there is one.
- Some farming areas in the UK have milk refill stations or offer milk in glass bottles in some local shops and farm shops.
- A Milksafe may be required in some areas to protect precious bottles. It's the only purpose-made milk-bottle protector that guards against cats, birds, frost and the heat of the Summer Sun. The optional lock will also deter opportunist thieves. It can be purchased at **Milksafes.co.uk.** There is a Milksafe to suit every household, and all sizes can be made to order, free standing or wall-mounted.

Milk – plant based in Tetra Paks

Tetra Pak, the Swedish-Swiss multinational, sustainability leaders, became the first company in the food and beverage industry to offer cardboard packaging made with a Sugarcane-waste, Bagasse, lining. The Sugarcane is grown sustainably in Brazil.

Some people now make their own plant milks and there is a lot of information online on how to successfully achieve this. If buying, try to source Bee-friendly plant-milks. Some Almond farmers use harmful pesticides that kill precious bees.

Mobiles/cell phones

There is no such thing as a plastic-free one at this point in time but there are companies making mobiles/cell-phones with replaceable parts. This is a good start. They are more expensive than the regular ones, but are built to last much longer and don't quickly end in landfill.

This is just one of the many plastic-dependent tech industries that needs a complete overhaul. Maybe Hemp plastic should also be explored.

Mobile/cell phone cases

- Pela makes sustainable cases that they claim are 100% biodegradable/garden compostable, but they don't cater to all phone brands at present, so check.
- You can send a worn-out Pela Case back for regrinding and reprocessing. They will also take back old mobile cases from any brand when a Pela Case is purchased.

Mobile/cell phone cleaning

Any microfibre cloth is usually plastic unless it states that it is a natural fibre. A soft cotton t-shirt rag dampened with a little ordinary soap and warm water is a far better cleaner, and won't damage screens the way harsh ones can. It can also be washed and reused with no waste. Just take care when cleaning near any of the outlets. Using a cotton rag with a little Lavender oil, that is anti-viral and antibacterial, is a good way of cleaning any type of phone, and smells lovely.

Modelling material/clay - The usual putty-like brightly-coloured modelling material is made from calcium salts, petroleum jelly and aliphatic acids. It has an unmistakable odour.

A Natural chemical-free alternative that also smells wonderful is Alkena Beeswax made in Switzerland. This modelling material is made from a combination of beeswax, natural fillers and earth colours. Softened by the warmth of the hands, it will harden slightly when cool. It can be re-used endlessly in the same way as the traditional material, with no harm to child, adult-modeller or Earth.

Mouthwashes - Most come in plastic bottles and, according to science, contain a nightmare cocktail of chemicals:

- Alcohol – makes bad breath even worse by drying out saliva.
- Fluoride – you probably already get a dose of this from your water supply and regular toothpaste and some tooth tabs.
- Benzalkonium Chloride – causes irritation of the mucous membranes, and skin around the mouth.
- Hexetidine/Oral Dene – has been known to cause blood clots in the brain, an irregular heartbeat, and allergic reactions, and is known to cause cancer.
- Methyl Salicylate – stronger than Aspirin.
- Methaparaben – a preservative. Parabens have been linked to breast cancer.
- Chlorhexidine – increases blood pressure. It kills all the good bacteria in the mouth and digestive tract that normally relax the blood vessels.

Do we really want to compromise our well-being with these toxic chemicals just to have fresh breath and a 'healthier' mouth?! All-natural mouthwashes are becoming the wise choice for the body and planet conscious. There are many recipes online.

Moving house

- Avoid Bubblewrap. Use cardboard boxes, tea-chests and even wine-boxes.
- Wrap delicate/breakable items in clothing, towels and newspaper. This is how it was done in the past.

N

Nail brushes

There is no need to go down the plastic route when there are Bamboo or Wooden ones available with natural bristles. At the end of their lives, they can be composted, put in a wood burner or recycled into woodchips at the waste tip.

Nail varnish/polish

Look for:

- Water and not oil based.
- Non-toxic.
- Mineral pigments and natural ingredients.
- Peelable. No remover required.

Nail varnish/polish reusable remover pads
These are available made of entirely natural materials like Bamboo, and Rayon made with plant materials (Cellulose).

Nappies/Diapers - plastic

These are global polluters, a waste nightmare, and a serious health hazard in countries with little or no waste management. They are often burnt, leading to highly toxic air. 20% of all river-pollution in Indonesia is used nappies/diapers. This needs to be addressed urgently!

Plastic nappies/diapers contain:

- Artificial fragrances.
- Chemicals.
- Plastics inside and as packaging.
- Adhesives and gels to absorb urine and faeces.

You'll never have to worry about running out of nappies/diapers again using these options:

- Cotton Terry towelling nappies and a lidded nappy bucket.
- Organic cotton nappies – where the cotton has been sustainably sourced.
- Bamboo nappies.
- Reusable nappy pants with a liner and poppers, made of Cotton or Bamboo. These are becoming increasingly popular and can be easily washed.

ReDyper, first launched in 2018 in the USA, offers home delivery of Bamboo nappies/diapers that don't contain any Chlorine, Latex, alcohol, perfumes, PVC, lotions, Tributyltin, or Phthalates. They're free of ink and don't have any patterns printed on them either. The American company Dyper has teamed up with

Terracycle and their industrial-composting partners to compost soiled ones safely. Subscribers send back used diapers in special bags and boxes that meet United Nations Hazardous Materials shipping standards.

Lots of information for new mums online.

Note pads

It's easy to make a green choice by purchasing recycled paper. Note pads and loose-leaf paper made from 100% post-consumer recycled content is what we are looking for.

If we always go for recycled paper, we can drastically reduce the number of trees cut down and processed into paper products worldwide. We can also use the back of printed sheets for notes and reminders.

O

Outerwear/coats and jackets

The Macintosh or 'Mac' was invented in 1823 when thick cotton was coated with a rubber solution to waterproof it. Also, in the past wool coats and jackets were the norm for Winter wear and cotton or linen for hotter times/climes. Wool coats were fabulously warm, heavy when wet and expensive and, sadly, that is why Polyester plastic ones came in. They are often made of materials coated in fluorinated chemicals that harm the environment. Some clothing manufacturers are hoping to do away with these chemicals in the not-too-distant future.

Look for:

- Wool - a superb natural alternative to plastic fibre that is cool in Summer, warm in Winter and hard-wearing.
- Organic cotton that is sustainably grown.
- Linen – made from the Flax plant.
- Hemp.
- Bamboo.
- Waxed cotton - waterproof and can be re-proofed.

Oven cleaners

Most are strong-smelling powerful chemicals that are not good for us or our pets to breathe in. There are more natural and safer ones on the market but you can just use good old reliable Bicarbonate of Soda.

Method:

- Make a paste with warm water, use a cotton rag to apply and then wipe off.
- You could use a coconut scrubber for stubborn areas but the Soda crystals are slightly abrasive and might be enough on their own without the risk of any damage to the lining.
- Ovens/oven-materials vary widely. Experiment with a small patch first.

Oven gloves/mitts - Avoid the ones made of plastic fibres, most are.

Use:

- Thick padded cotton that will last for years.
- Knitted natural string ones that have been popular for decades and will last many years.

Try to also avoid Silicone ones that are not biodegradable at the end of their lives because they are often made with mixed materials that include plastic.

P

Padded envelopes

- This type of mailer is usually lined with bubble-plastic and is sometimes completely made of plastic.
- You can get eco-friendly paper ones that are padded with recycled shredded paper. How they used to be made.

Paints for children, and hobbies

We often aren't aware that acrylic paints are in fact plastic and contain nasty chemicals that get into waterways when brushes and containers are washed out after use.

Watercolours are made of natural minerals and are readily available from art and craft businesses. They can be bought as powders in tins as well as in liquid form in metal tubes.

Paints for decorating

Polyurethane, that most paints are made from, is a plastic-based resin, and is used to manufacture anything from furniture to baby toys because it is hardwearing. We need to avoid these now and use non-toxic paints that are water based and will not harm life.

Lakelandpaints.co.uk produce a range of paints that are odourless, organic, non-toxic, VOC (Volatile Organic Compounds) and solvent free, as well as being plastic free. They are also suitable for the Nursery and for painting toys.

Paper kitchen towels

Use instead:

- Rags and hot water or steam – cotton is best
- Old knickers and underpants
- Old shirts
- Old towels
- Worn through flannelette bedding
- Face flannels

Just leave one of these rags handy by the sink and keep it rinsed out with hot water for cleaning surface spills. I use an old cut-up terry towelling bathrobe and very hot water for this purpose, and also for cleaning the front of appliances and the stove. When a rag gets manky it can go in the compost because no chemicals have been used, only water.

'Un-paper towels' on a roll made of recycled cotton materials can be bought from artisan makers. Or make your own with layers of flannel. Be creative with your designs and have fun!

Paper tissues

Usually come in plastic and should be avoided altogether. A phenomenal number of trees around the world are cut down just to make them and, in some cases, the cardboard box they come in. Old-fashioned cotton hankies rock for a quick nose-wipe and you can throw them in the wash. They can be found in charity shops, or ask grandparents if they have any spare. Hankies will last for decades if you look after them and then can be composted. I'm still using some I bought 40 years ago in India! I know some people think they're gross but if you use one a day and throw it in the wash, I can't see the problem. You'll never suffer from the tissue shredding in the washing machine scenario!

Use loo roll instead of paper tissues or cotton hankies for colds and flu and then burn them safely if you can.

DIY Hankies can be upcycled from cotton:

- Shirts and blouses
- Nighties
- Pillowcases
- Bedsheets/bedding

We all need choices, and LastTissue™ from the Last Object company has developed a clever modern take on the cotton hankie that comes in a handy Silicone dispenser. They say: 'Use, wash, store, repeat.'

Party planning

We should never think Plastic first!

We could ask each guest to kindly bring their own kit, no single use plastic allowed, thank you, with a bag to put the dirties in.

Include:

- Plate
- Cup/mug/beaker
- Bowl
- Cutlery
- Washable napkin

Don't stress if your guests turn up with hard durable plastic. If it's already in the house it should be used until it dies and then be replaced with something more planet-friendly.

Maybe if you know your guests well you could ask them to bring some food too for a pot luck meal. A shared meal with good friends is one of the joys of life! No washing up or clearing up for the host/s. Win all round!

Natural party bags for children to avoid the plastic tat. There are a lot of online companies that specialise in natural Eco party bags but you can do it yourself.

You might like the following suggestions:

- A colourful book on Nature or the effect of plastic on the planet – see **Recommended Books** page.

- Wildflower seeds or Sunflower seeds, and all they need to plant them, in a paper envelope.
- Watercolour pencils and a sketchbook or colouring-in book, double the fun!
- A wooden Yoyo and other small wooden toys.
- A cotton fabric bag to decorate along with some fabric crayons.
- A Cactus or Succulent, without any spines – easy to look after. Some species even flower. It might start a lifelong passion!

Pass the parcel can be done in a natural way.
Colourful cotton drawstring bags, diminishing in size, can be used instead of wasteful paper parcels. They can be washed and reused time and time again and passed down the family. You can buy or hire or make your own unique designs.

Pegs for the washing line

Wooden pegs used to last a lifetime, some of mine are probably 60 years old, but the modern ones in the last few years have been really poor quality.

Look for:

- Good quality wooden pegs with strong springs from a sustainable source: FSC mark (Forest Sustainability Council)
- Bamboo
- Stainless-steel Cyclone pegs

Pencils

- We should avoid the ones that are retracting and in a plastic casing if possible.
- Some wooden pencils have seeds embedded in them to plant out at the end of their life. They make great gifts and can be put into 'natural' party bags.

Eco Happy Pencils in India make pencils from recycled newspaper that cuts out the need for wood. The majority of their staff is female, giving local women the opportunity to earn a decent wage in safe working conditions.

Stocked in the UK by:
The Plastic Free Shop
Peace with the Wild

Pens for personal use

These everyday items usually end in landfill, contributing to the massive amount of plastic waste poisoning Earth. Improperly disposed of, they break down into microplastics over time and pollute our global rivers and oceans.

A great alternative to the ubiquitous Biro type pen is an elegant metal Fountain Pen. A lot of people prefer them and you can get different colours of ink for them. The ink is syphoned into a chamber from an ink-bottle, which is usually glass. Fountain Pens are not so tiring to use either as you don't need to press down hard to make them function well.

Another alternative is a Cartridge Pen, where cartridges slot into the pen and can be easily replaced. The cartridges are plastic, though, and have to be binned when empty.

Both of these options can last a lifetime if properly used, and were common items when I was growing up. I hope a home compostable cartridge for the latter can be developed soon.

Disposables

If we are going to continue down this path, any pen must completely disappear and go back into the earth with no harm to life. You can get biodegradable pens made of wheat grass now that will do just this.

Handcrafted Artisan-made Glass Pens are becoming popular with calligraphers and graphic artists.

Periods – sanitary wear

Sanitary wear contributes enormously to global plastic pollution because it involves plastic somewhere.

Sanitary pads that have a plastic lining also contain toxins that can disrupt hormones. Interfering with hormones can cause health problems, including lessening resistance to disease.

Tampons, that were invented in 1933 in the USA, are even worse because they usually include a plastic applicator and go inside the body, where there's the real risk of Toxic Shock Syndrome. We now know that plastic is extremely bad for health so we don't want all those nasty chemicals and plastics next to the most delicate parts of the anatomy!

The usual plastic type of sanitary wear contains:

- Artificial fragrances.
- A cocktail of chemicals.
- Plastics inside and as packaging.
- Adhesives and gels to absorb blood, or urine in the case of incontinence pads. (Natural plastic free incontinence pads are available in many countries too.)

Pads, with their plastic anti-leak linings are dumped everywhere. Tampons and plastic applicators end up in waterways and oceans worldwide.

Never flush away a pad, panty-liner, incontinence pad, tampon or tampon applicator! Even if the applicator is carboard, it was never designed to go down a loo. It will not magically disintegrate but will block pipes!

Also, never flush away unwanted pills, that will contaminate and poison our river-systems and oceans. Take them back to a chemists or pharmacy so they can be responsibly disposed of without harm to the environment.

Tampons are the 5th most common item found on Europe's beaches during regular beach cleans.

These are the alternatives:

- Cotton reusable cloth pads with poppers – lots of patterns to use online if you want to make your own.

- Organic Cotton or Bamboo reusable cloth pads that you can buy – some have a charcoal layer that absorbs moisture in place of synthetic gel.
- Cotton period pants - which can be worn for up to eight hours at a time and then washed.
- Silicone period/menstrual cups that can last 10 years – there is a lot of information about their use online.
- Plastic free tampons, with carboard applicators.

Advice on natural periods can be found online, as it is no longer a taboo subject in most Western societies. Many artisan makers are changing the world of menstruation for good and a few major supermarkets have realised that this is the way the public wants it now. Natural is the only way forward!

Pets

Don't despair. The wish to be plastic free and healthier is slowly but surely trickling down to our beloved pets. There are pet toys, collars, leads and beds made of natural materials now.

Pet collars and leads

To avoid both leather and plastic, these are alternatives:

- Hemp corduroy
- Plaited cotton
- Harris tweed
- Cork – a lovely soft and sustainable fabric from the Cork Oak. There may be a plastic fastener

but it is early days for this sort of collar or lead so one small bit of plastic is better than 100% plastic.

Pet shampoo is available in bars and is specifically formulated for pets.

Pet toys, bowls and bedding

- Dog and cat toys made from natural rubber or corn starch can be sourced. You can also get play balls for cats that sometimes have a metal ball inside but, as always, do research and make sure they are safe for your pet.
- Avoid hard plastic and polyester beds and bedding if possible and source natural materials. You don't want your precious fur babies inhaling or ingesting plastic fibres. The traditional wicker pet baskets are still around. You can also buy beautiful colourful woven beds ethically and sustainably produced in Ghana from Veta Vera Grasses from Thebasketroom.com.

If you buy petfood/biscuits loose at a pet store, take your own container/s to avoid the use of plastic bags. It is an easy habit to get into.

Birds

- Source feeding/drinking bowls in safe metals such as stainless-steel if they fit the cage. Sometimes, sadly, you have to put up with the plastic ones that come with the cage if buying new.

- You can buy metal forage-balls to put treats and millet-sprays into.
- Take your own containers for seed and millet if you buy from the bins instore to avoid any plastic bags if you only need small amounts.

Rabbits, guinea pigs and other small pets

- Use stainless-steel bowls and water fountains.
- Source plastic free toys if you can. Give them as natural an environment as possible.

Fish tanks

Ceramic ornaments used to be readily available for fish-tanks, and, fortunately, some pet-shops still stock them. We must avoid putting anything plastic in the tank now, including greenery or the usual plastic diver. Plastic is not good in a watery environment and can pick up bacteria which will stick to its surface. Be safe and not sorry with your precious fish and other water life! Also, we shouldn't buy corals or shells. They should be left in our oceans where they belong.

Picnics and food on the go

We can take our own:

- Reusable water-bottle – avoid the plastic ones, stainless-steel is best but some people use glass or Bamboo. Make sure the Bamboo water-bottle is a good one from a reputable company.
- Cup – stainless-steel, enamel or Bamboo. There are Silicone cups but the jury is out on the

safety and end of life disposal of that particular material.

- There are collapsible ones in both stainless-steel or Silicone.
- Lunchbox – stainless-steel is best but some are even made of wheat grass now. (I have a wheat grass one and it is almost indistinguishable from plastic and does exactly the same job.)
- Cutlery and chopsticks – stainless-steel, Bamboo or wooden are best.

Take any wooden or Bamboo cutlery home from eateries to wash and reuse for picnics. Please don't bin them straightaway. Give them a long life!

- Stainless-steel camping trays can replace the usual plastic plates and are much easier to eat from as they are divided into 2 or 3 compartments for different foods. They are easy to clean and store, and will last decades.
- Straw – stainless-steel, stainless-steel with a Silicone tip, paper or Bamboo. Even hollow pasta will make a straw!
- Chopsticks – wood, Bamboo or stainless-steel.
- Spork – a handy neat combination of a spoon and a fork that can be in either stainless-steel or Bamboo. They are sometimes found in a handy soft Cork pouch.

Edibles

- There is a Mexican company that makes cutlery out of Avocado Seeds.

- There is also an Indian one that makes cutlery out of Edible Biscuit.
- Others make edible crockery and cutlery out of plant materials.

In future we may not just be eating the meal but the crockery and cutlery too, if we so desire! I think I'll just compost or worm bin it!

Pipe-cleaners for crafting

Cotton and wire ones are available from specialist craft suppliers.

You don't need to go down the plastic route that most people think is the only choice.

Planters/plant pots for the garden and house

We shouldn't just throw out any plastic. We need to use it until it dies before replacing it with something more durable and non-plastic.

We can avoid the plastic pots/planters, and old tyres that leach toxins, by using:

- Ceramics – frost-proof of course for any outside space.
- Terracotta – gives a warm Mediterranean vibe.
- Clay fibre/Fibre clay – used for pots - made in moulds, with a resin mesh or fibrous substance with minerals such as stone or clay. It makes an attractive and lightweight substitute.

- Concrete – does crumble eventually but can last decades.
- Old sinks – a Butler, deep sink is perfect for Alpine plants or a pond for water-plants. Put in something to help wildlife climb out if they accidentally fall in if you do make a pond.
- Old metal birdcages make lovely ferneries, even when rusty!
- Homemade brick planters – lots of videos online.
- Anything that can be planted into, like ceramic teapots and cups, even old leather shoes! As long as the plants have enough room and get good light, they can be planted into almost any vessel.
- Reclamation yards for something unusual and quirky to add an element of surprise to your outside space.

We must also try to avoid buying any plastic-handled tools, statues, gnomes, and windchimes/toys, etc., things that are often overlooked when going plastic free in the garden. Lovely Bamboo and metal windchimes are available.

Avoid buying any inflatables if you can as there are only 3 options for disposal:

- Landfill – most end in landfill.
- Incineration.
- Send to a company that makes bags out of old inflatables and pool toys, like Wyatt and Jack. Better than landfill.

Traditional plant mist-sprayers come in copper or brass so we don't need to use plastic, and metal watering-cans for outside and indoor use can still be found. Just keep the garden ones in a shed or under cover to prolong their lives.

Plasters/Band aids

- 'Patch' plasters are plastic free and made of Organic Bamboo Fibre with a Coconut Oil Gauze.
- Some have Activated Charcoal in them to help heal infections and are good for sports enthusiasts and children.
- They are natural, don't leave marks or rashes and will not irritate sensitive skin.
- Chemists, Plastic free/Zero waste shops and online businesses often stock them.
- They come in a cardboard tube that is home compostable, and the plasters are home compostable after use too.

Plastic food containers

- We should never microwave polycarbonate plastic food containers. Polycarbonate is strong and durable but over time may break down at high temperatures and the plastic toxins may leach into food.
- Some plastics that are marked with recycle codes 3 or 7 on the bottom may be made with Bisphenol-A or BPA, known to be a hormone-disrupting chemical.
- **Bisphenol makes plastic hard when needed.**
- **Phthalates make plastic soft, as in film and packaging.**

Both have been found in all the human urine samples in studies carried out by Dutch scientists. We all carry them in our bodies now and only goodness knows what they are doing to us!

What we can do to avoid them:

- Reduce our use of tinned food – most tins are lined with this plastic. Some food companies are now working to find a non-plastic alternative to the lining that prevents the metal from corroding.
- Buy food in glass jars.
- Cook for ourselves using fresh ingredients.
- Reduce or cut out our use of drinks cans.
- We should opt for glass, porcelain or stainless-steel containers for hot food and liquids.
- Source baby bottles that are glass with a natural rubber teat.
- We shouldn't store food in plastic but use glass when we can.
- Say no to till and ATM receipts that often have a coating of plastic on them. Only keeping the supermarket/store ones we need for proof of purchase in case of return.

Play tents and tepees for children

In the past an old sheet thrown over the washing-line and pegged down either side made the best temporary tent!

A wonderful, natural living alternative is to make one out of Willow wands if you have outside space. Beats all the plastic and will last a lifetime. Nurseries selling Willow and instructions for using Willow wands and for making living shelters can be found online.

For the gardeners among us, you can even make a simple garden tepee out of Runner Beans. They have lovely flowers too and will make a pretty temporary tepee.

All you need for your Runner Bean Tepee:

- A nice patch of fertile stone-free soil.
- 10 or 12 strong Bamboo canes or tree branches at least 6 feet/about 2 metres long pushed into the soil securely in a large circle.
- Some twine to tie the sticks together in a tepee shape at the top.
- Chicken-wire or a web of gardening string around the poles, leaving an entrance.
- A seedling by each stick when they are in place. Wind the plants around the sticks to give them a good start.
- Water for a thorough and regular soak.
- Patience to watch them climb and wind their way up the poles.
- Children who love Nature and dirt!

Slugs could be a problem but we shouldn't use any chemical slug pellets, those nasty blue ones, that can harm/kill other wildlife or domestic pets. Protect seedlings from slugs with organic slug pellets, beer-traps or by removing them each evening until the bean plants are strong enough to be an unattractive prospect to any hungry slug. The beans can be started off in pots and then transplanted when they are 6 inches/15cm tall.

Plughole cleaning if there's a pong

Avoid all the nasty chemicals in plastic with 1 cup of Bicarbonate of Soda - can be sourced in cardboard in the UK.

Method:
Mix Bicarbonate of Soda in a jug with hot water from the tap and slowly pour down the drain to freshen.

Plughole cleaning if there's a blockage in the shower

Most blockages are caused by entangled hair and soap.

Use one of these to remove the hair first and bin the mess:

- Pair of tweezers
- Wire coat-hanger hook
- Chopstick
- Anything that can grip onto the hair and that you can pull. I use an old dish-washing brush.
- There may be a trap that you can unscrew and clean.

We don't need chemicals in plastic if there is a blockage in the bathroom or kitchen sink. Keep a Rubber Plunger under the kitchen sink. They only cost a little and could save you a fortune when you don't need to call out a plumber. Get a real rubber one, that gives good suction, and not a plastic one that is not as effective, and we know is bad for the planet anyway.

Bicarbonate of Soda – that can be sourced in a cardboard box in the UK – and Vinegar is good for blockages.

- Run hot water down the drain to dissolve any fatty deposits.
- Put in half a cup of Soda.
- Wait about 15 minutes.
- Mix a cup of vinegar with warm water and pour that in too. It will sanitise the drain.
- Run the hot tap to clear any lingering smell.

Polystyrene/Styrofoam food-containers

- We must avoid them altogether and look for food outlets that use boxes and containers made from Sugarcane waste, called *Bagasse, or sustainably-sourced cardboard.
- You could ask them to use your own containers. It is always worth a try.
- Buy a set of Tiffin Carriers, see **Takeout food** for more details.

*Products made of Bagasse are compostable and manufactured in a high-heat, high-pressure process. They can be used for hot drinks, hot, wet or oily foods and are sturdy and economic, costing little more than Polystyrene to produce.

The usual Polystyrene packing peanuts are horrible toxic things that should be banned worldwide. These ultra-light bits of plastic get everywhere on a windy day, then embed themselves into the ground or float on water to pollute and poison life for decades.

If you receive goods from plastic free shops and it looks like they have been packed in polystyrene peanuts, take one to see if it will dissolve in water. If it does, then it is made of corn or potato starch and is completely safe being composted or washed down the drain. Personally, I would compost them or give them to my worms.

Poppy Day for The Royal British Legion, UK

- You can avoid the paper and plastic poppy if you want to by buying a charming brooch or pin from their site, wearing it each year and donating. They make lovely thoughtful gifts too.
- Some people crochet or knit their poppies and donate. Just avoid the plastic fibre and use cotton instead.
- You could make some from paper or card if you are feeling arty and donate.

Puppy pee pads

- Please avoid them altogether when training your puppy.
- They are more or less giant disposable sanitary towels with a plastic backing.
- Some pet owners claim they can actually do more harm than good in the long run as it teaches puppies that it's all right to do their business indoors.

  An eco-friendly biodegradable alternative is a real grass pad that is becoming popular with owners who have no outside space. It comes in several sizes to suit the size of your dog.
- It arrives in strong cardboard and is soil free and mess free.

- One company that supplies these in the UK is ‘Piddle Patch’. There are others in the USA, one is ‘Fresh Patch’.

Q is for quality and not quantity. We should always look for the best we can afford at the time that is going to be a long-lasting product so we aren’t forced to buy and keep buying. The problem in the last few decades has been that the quality of most everyday items is exceptionally poor, and that these products have to be discarded soon after purchase. The planet and our pockets can no longer afford this sort of waste!

Q-tips – See cotton swabs/buds

R

Razors - to replace disposable ones

- Stainless-steel Safety Razor – unscrew the head to change blades.
- Stainless-steel Butterfly Razor – turn the handle to open the top to change blades.
- Pivoting Head stainless-steel safety razor - unscrew the head to change blades.
- Bamboo-handled metal razor – unscrew the head to change blades, and keep it dry between use.
- Complete shaving kits are also available.

Using a safety razor:

- Apply plenty of soap.
- Go slowly to avoid nicks.

- Dispose of blunt blades by collecting them in an old tin and then put the whole lot into the metal recycling bin.

Ready meals - very handy for busy people but we should try to avoid plastic trays, microwaveable trays, and plastic film when we can. No plastic is good for our health and especially when heated.

- Buy from a butcher and fishmonger, if possible, that wraps in paper.
- Buy fresh produce whenever possible.
- Go for meals in cardboard or cardboard and foil. If you eat meat, aim for a couple of meat-free and fish-free meals a week to help the planet.
- There is a large selection of plastic free veggies and delicious veggie/Vegan meals in most supermarkets and independent stores that come in carboard in the freezer section.

Doing the research for this book taught me so much about sustainability. I could no longer watch fish stocks dwindling from over-fishing or the number of plastic particles in fish escalating. So, I decided to finally give fish the push in my life altogether. I'm now 100% Veggie. I feel so much healthier, and have played my miniscule but vital part in starting to heal our sweet Earth.

Receipts – tills and ATMs

We can refuse receipts unless we need them for proof of purchase as the shiny ones contain an alleged hormone-disrupting chemical called Bisphenol-A (BPA). It's an easy habit to get into and we won't have drawers full of the bloomin' things either!

Red nose Day for Comic Relief UK

The good news is that the famous red noses, that used to be littered everywhere, are going to be made of, good old Sugarcane waste again, Bagasse from 2021. I'm not certain how these will break down in the environment, and only time will tell as they are a hard bioplastic.

We can make our own fun noses and donate.

- Make a paper one.
- Make a wool one.
- Make a felt one.
- Paint your nose red, or with a character, for the day with non-toxic face paint.

Ribbon

- Most is plastic now.
- You can still buy real Silk ribbon from specialist high street shops and online stores if you need a bit of luxury.
- Raffia, from the Raffia Palm tree, that is grown especially in Madagascar, and Central and South America, makes a lovely natural alternative to the usual shiny plastic ribbon and comes in naturally dyed colours for a bit of added style.

Rubber/plastic duck bath toy

Real latex rubber ones are available for your children's bath-time fun that are biodegradable. The plastic ducks used in races are usually collected but I bet there are a few escapees still bobbing about in rivers endangering and frightening the wildlife! Maybe we could also persuade the organisers to go for rubber duckies instead!

Rubber gloves for household jobs

- Most 'rubber gloves' are in fact plastic.
- You can still get real latex rubber gloves, if you aren't allergic to them, that will biodegrade in a home composter or naturally in the environment.
- 'If You Care' uses Fair Rubber to make their gloves that are lightly dusted with cotton inside.
- They recommend using one pair for dishwashing and a second for other household cleaning/ chores.

S

Sachets/plastic pouches – whatever type and use

We need to void these altogether as they are one of the planet's biggest marine-polluters and need to be banned worldwide. These mini plastic pouches do give some of the poorest people in Africa and Asia access to everyday household essentials that they normally wouldn't be able to afford. Sadly, it is also a way for the multinationals that manufacture them to increase sales by targeting the poor. This is a despicable practice!

Sixty billion sachets a year are sold in The Philippines – from a study carried out by The Global Alliance for Incinerator Alternatives (GAIA). Altogether the global count is 855 billion, according to A Plastic Planet. This has to stop as the general pollution is phenomenal. We must see an end to petrochemical plastic or the planet will never recover!

There are other ways of doling out small amounts that don't involve single use plastic, such as products dispensed into refillable bottles from bulk containers. These companies can do it if they really want to. They have the money and the clout to end the sachet culture or find alternatives like seaweed and algae that will biodegrade in water. These new bioplastics are being developed right now but it's early days yet for anything commercially viable. Time will tell.

Sauces

We should buy these in glass bottles or jars to avoid the 'squeezy' plastic ones that are usually more expensive anyway.

Sellotape

- After years of being made of plastic, this has gone back to the original Cellulose, plant material, and comes in recyclable cardboard packaging with a cardboard ring/core. Thank you, Sellotape!
- Another eco-friendly alternative to any plastic tape is paper tape that is either already glued or needs to be damp in order to stick. Source this from stationers, plastic free/zero waste shops and online businesses.

Sequins

We shouldn't buy/use them at this time as they are made of plastic. If the sequins shed off garments, cushions and bags, and enter the environment, they will never go away and could be ingested by animals, birds, fish and other aquatic life. Years ago, they were made of thin metal but were heavy, expensive and took huge resources to make, and that is why cheaper lighter plastic came in.

A clever scientist has recently created shiny sequins made of cellulose, plant material, that are compostable. Even if they do get into the environment, they will not harm wildlife or sea life because they will break down into organic particles and disappear altogether. I hope these become commercially available soon. Manmade products need to go back into the earth without harm to any living creature or Mother Earth Herself.

Shampoo in plastic bottles or sachets

- A Shampoo bar is one of the best alternatives.
- There is now one for every type of hair and water-condition. It may take a while to find the perfect one.
- They're also great for travelling as they are not subject to the 'liquids rule' at airports.
- As hair is stripped of oils by more commercial chemical-heavy shampoos, you may find your hair feels different after using a shampoo bar made of natural ingredients. It may not sound squeaky-clean after washing but will still look and feel clean when dry.

- With some bars, you may also need a rinse of Bicarbonate of Soda, Apple Cider Vinegar or Lemon but others only need a thorough warm water rinse.

Experiment on days you don't need to look your best. Eventually your hair will settle down and you won't be washing it so often as your hair's natural oils take over.

Finding a good bar for your hair type means you will never go back to the bottle.

I switched to a bar shampoo, love it, and will never go back! My greasy hair stays cleaner for longer, at least 5 days now instead of 3, and there's no plastic or nasty chemicals anywhere! Go on, give it a try!

Using shampoo bars:

- Wet the bar and hair and rub the bar all over the head and massage in. It's a bit odd at first but you will get used to it.
- Rinse thoroughly and repeat if necessary.
- Some bars are so effective that a repeat application is not necessary. The most important thing is to rinse the hair well so no residue is left.

Some people use the 'No-poo method' where they don't use anything at all but warm water. The scalp 'learns' to balance the amount of sebum produced over time. It works for some and not for others so it might be worth experimenting with. There are support groups on Facebook as it is quite a complicated subject, there being so many different types of hair, and water conditions, around the world.

EarthSuds are shampoo, conditioner and body-wash cubes that can be used at home but could also solve the problem of single use plastic bottles and sachets in the hotel industry. They were invented in 2017 during a competition for circular design solutions by the 19-year-old Canadian Marissa Vettoretti.

Use organic **Dry Shampoos**, which have no nasties or plastic in them, to save water and washing greasy hair so often. They are a bit expensive for everyday use but great for holidays, festivals, and bad hair days. There are recipes online for DIY versions that some people swear by.

Refilling is also becoming popular for shampoos and eco alternatives to many chemical-laden products, and is a service that can often be found in zero waste/plastic free shops/stores.

Shaving

- If you use a good electric shaver then carry on. (Mine is over 40 years old and still going strong!)
- If you like to wet shave then avoid the plastic razors and invest in a good stainless-steel forever razor. There is a huge variety now. **See Razors,** for types.

Gentlemen: You can also find natural shaving soaps but will probably have to get a non-natural/Vegan shaving brush as the only alternative is the old-fashioned badger-hair one that is rarely made now, rightly so, because of animal concerns.

All-natural hair and beard care products are available plastic free too.

Shoes/trainers - See Footwear

Shoppers/Shopping bags

Take your own reusable ones – see **General Advice**

Shower curtains

- Most are made of plastic.
- The Organic Shower Curtain Company has a selection made from waxed linen or organic cotton.

Shower head cleaning

- Use 1 cup of Vinegar (can be sourced in a glass bottle).
- Pour Vinegar into a flat bowl and immerse shower head.
- Leave for 30 minutes.
- Rinse to remove limescale and the strong smell.
- Apple Cider Vinegar is just as good but without the strong pong and can also be sourced in glass.

Shower gel

Look for natural ingredients in compostable corn starch sachets.

Silicone products

Silicon is extracted from silica sand, that glass is made from, and passed through hydrocarbons, organic compounds that occur in petroleum, natural gas and coal, to create Silicone, with an 'e'.

Silicone is subsequently not as planet-friendly as we are often led to believe. It's just another product of Big Oil. As it is a manmade substance, it is not biodegradable. It's also more resistant to heat, used for kitchen utensils and baking, and may take longer to break down than plastic if it ends in landfill. It can't normally be recycled either. The only good thing about it is that products made of Silicone usually last decades and avoid the use of single use plastic.

I'm not a big fan of anything made of the material. To me and many others, it's a case of swapping one type of plastic for another.

Sledges for snowy fun

We should be buying a good one that will last years. Not a flimsy plastic one that will break first time out, and are often found abandoned after a snowy weekend. Strong well-crafted wooden ones can still be bought from artisan makers.

Slug control in the garden

We should never use chemical slug pellets in plastic or anything else! They're highly toxic to all life, especially to animals like our wonderful Hedgehogs,

Frogs, Toads, Grass snakes and Slow worms whose diet includes Slugs.

Protect seedlings from slugs with:

- Organic slug pellets that don't harm life.
- Beer-traps.
- A sprinkling of crushed eggshells as a deterrent.
- Copper tape on plant pots that contain a tasty meal for them.
- Go out with a torch and remove slugs each evening until plants are strong enough to be an unattractive prospect to any hungry slug.
- Tiny slugs on new growth are the main problem. The bigger slugs don't do as much damage and some are slug-killers themselves.

We must stop using pesticides/biocides/chemicals and employ a more natural way of growing plants, especially for food. Companion planting, growing plants that attract pest-predators, is the way of the future wherever we live on the planet. Please start now!

Snacks and nuts

- These are hard to find in compostable packaging at present but Snact.co.uk use compostable bags made by the company Tipa.
- Being as durable and impermeable as ordinary plastic, it is safe to use for packaging food.
- The bioplastic film breaks down in a home composter within 6 months and becomes fertiliser for the soil.

- Nothing will change unless we tell companies how we feel about their plastic packaging. Let them know that others are stealing a march on them with their home compostables. We can't subject our glorious planet to any more waste!

Soap dishes

Now we're using bar soap we need to go back to some sort of drainer for our solid soaps, shampoos and conditioners. (I have a turtle-shaped one made of Corn starch, as I love those wonderful creatures.)

We could use:

- A cut-up piece of loofah to put the bar onto, but this must be kept in a moisture free environment or mould will be a problem.
- A slatted wooden drainer – a large variety available.
- A ceramic drainer with holes – artistic ones available.
- A coconut scrubber to place the bar onto to dry – popular.
- A net or bag hung somewhere to dry.
- A bottle cap stuck to the underside of the soap to lift it up out of the wet, left somewhere warm to dry.

Socks

Most are made of polyester or mixed fibres that may include natural materials but are still no good for Earth.

Look for:

- Wool – warm in Winter and cool in Summer.
- Organic cotton – comfortable and cool.
- Bamboo - makes a soft fabric that lets your feet breathe and has strong antibacterial and antifungal properties to help keep them healthy.

Soft toys – commercial – See Toys

Soft toys – homemade

- Polystyrene pellets/balls should not be used for any toy or bean bag making as they can get into the environment if the article is torn.
- Bigbeanbagcompany.com in the UK makes a Polystyrene-bead alternative from a foam made of eco-friendly plant-based materials.
- You can avoid the Polyester stuffing by using real natural Kapok, cotton-like fluff obtained from the seed-pods of the Kapok tree.
- Cut up old cotton t-shirts can be used for stuffing too, so the item can be washed.

Sponges for household cleaning

Look for sponges made of plant cellulose which can then be composted if you don't use any chemicals, that includes Lemon, Vinegar, Citric Acid or Bicarbonate of Soda. Just use them with hot water. They usually come in a cardboard sleeve.

Spray bottles for homemade lotions, potions

Some people reuse plastic ones after a thorough clean but the following are plastic free alternatives:

- Glass bottles with a metal spray cap - becoming very popular now.
- Aluminium bottles with a plastic spray cap - lightweight, could last a lifetime, and aluminium can be recycled.

Storage

This is a tricky subject when it comes to plastic as we are so used to all those huge plastic boxes that are so handy.

Keep some cedar blocks in these alternatives to deter moths:

- Seagrass boxes – expensive but rather charming.
- Wooden chests – you can get flat-packed ones.
- Rattan boxes – from Oriental suppliers.
- Bamboo boxes - from Oriental suppliers.
- Metal ant-proof trunks used for travelling – you can buy them from specialist luggage shops. They look very attractive and do a good job. (I loved mine but had to leave it in India when I came home.)

Storage in the kitchen

- Locking glass Mason jars – becoming popular again and can be found inexpensively.
- Spare jars of all sizes – you may need a stash to go bulk shopping.

- Ceramic pots with lids – lots in charity/thrift shops.
- Old biscuit tins for cakes and biscuits – save the Xmas ones!
- Glass casserole dishes with lids for fridge storage – you can now get reasonably priced locking ones with a plastic or Silicone seal that is better for health than having something entirely made of plastic. Old fashioned glass-lidded casserole dishes turn up in charity shops, or ask Granny! (I still have some that came from my dear Granny and Mum. They must be half a century old now, and are just as handy as ever.)
- Glass can be used in the freezer but you must leave an air-pocket at the top of any jar or container so the glass doesn't explode. I'm told it is quite an art to get it right. Check the online blogs/advice first before trying.

Straws

The only thing that matters is what they are made of. I'm not going to discuss the yes or no of straws and am just thankful when they are not plastic and helping to destroy the planet. In Italy some bars and restaurants are even using pasta! If we are going to buy straws for our homes, when there are no medical requirements, we should be thinking of forever ones made of stainless-steel.

When out and about look for:

- Paper.
- Stainless-steel – will last a lifetime.
- Bamboo – keep it dry between uses.

- Silicone tipped stainless-steel – might not last as long as a fully stainless-steel one if used on a regular basis.
- Real straw – how straws got the name in the first place.
- Glass - not practical in some situations but you can buy strong Pyrex-type ones.
- Telescopic stainless-steel straws that are scratch proof, dishwasher-safe and environmentally friendly. They come in a box or on a key-chain, or fold down in some way for ease of carrying. Some also come with a brush for cleaning but it is likely to be plastic.
- The Sulapac Straw is made from sustainably sourced raw materials, and won't go soggy. It can be organically recycled, and the wood it contains comes from wood that would otherwise be wasted. If it accidentally ends up in an ocean, it won't harm the delicate ecosystem and will naturally biodegrade.

Sugar

- Most White sugar sold in the UK is in paper packets.
- Brown and Golden sugar usually comes in plastic.
- You can make your own Brown sugar from Treacle or Molasses which comes in either a tin or glass jar.

Method:

- Mix 1 tablespoon of Treacle/Molasses to one cup of White sugar.
- You may need to experiment a bit to taste.

It's advisable to keep any sugars in glass mason jars because paper packets go soggy and deteriorate in damp conditions. Even though the sugar can still be used, you risk dropping the damp packet on the floor! So, decant any sugars in paper into snap-lid jars and then recycle the paper while it is still clean and dry.

Sunglasses

- Wooden, and Bamboo, framed-ones are available.
- Those made from UK wood are understandably expensive.
- Sunglasses made abroad are often a lot cheaper but we have to weigh up the pros and cons before buying.
- One company is constructing them entirely from plant-based bioplastic that will decompose in the environment.
- Some are now made of recycled cotton, wood, and paper.

So plastic no longer needs to be our first and only choice.

Sunscreen/Sunblock

Found in tins or glass:

- Zinc
- Minerals
- Natural creams

Make sure you buy Ocean/Coral Reef/Environmentally-friendly ones so they don't damage life below the waves and add to ocean pollution when you go swimming.

These can be found in plastic free and zero waste shops and online businesses that care about all things Eco.

In 2019 The Pacific nation of Palau became the first country to ban sunscreens that are harmful to corals and sea life. Sunscreens that include ingredients such as Oxybenzone are no longer allowed to be worn or sold in the country. A recent scientific study has found that the common ingredients of sunscreen can also be absorbed into our bloodstream. Scientists need to discover whether they are having a negative impact on our health as well as that of the wider environment.

Sweets/candies

Plastic-free confectionary is hard to find in most places but can be sourced from specialist sites and shops. These more traditional sweets are often palm oil free and Vegan too. Vast Palm oil plantations are causing widescale deforestation and the destruction of whole ecosystems.

Look for sweets/candies that come in:

- Paper
- Paper and foil
- Tins
- Carboard tubs/boxes/tubes that can be recycled or composted.
- Real Cellophane that is plant based and therefore biodegradable/home compostable.

Make your own sweets

There are some lovely recipes online and in specialist books for the boiled variety, or money-saving fridge-type chocolate bars that are simple to make with just a few ingredients. These make good picnic/lunchbox treats.

Swimwear for water sports

Yulex is a material made from the natural rubber of the Hevea Brasiliensis tree that is being used to make wetsuits and watersports-wear to replace the usual Neoprene and plastic. It is FSC Certified by the Rainforest Alliance.

T

Takeout food

Take your own containers if you can as well as a reusable bag to put them in. It will be completely mainstream soon and you won't need to feel odd doing it. You can also take a doggie bag or container to a restaurant for leftovers.

Also, remember:

- Refuse the plastic sachets!
- Refuse the Soy Plastic Fish that many Asian takeaways supply. These are a huge problem in Asian rivers, and are now turning up as pollutants around the rest of the world.

Make a to-go kit that you can keep in your car for those quick dashes that includes:

- A cloth napkin.
- Stainless-steel or Bamboo cutlery.
- A reusable coffee-cup.

- A good non-plastic water-bottle, ideally stainless-steel.
- A couple of different-sized containers with lids.
- A generously-sized tote bag.

Some Indian takeaways now use Tiffin Carriers, a wonderful Indian invention made of a stack of light stainless-steel containers, locked together. These have been used for well over a hundred years to ferry around lunches from houses to workplaces. They are still used daily in India in a highly organised and very impressive way.

Tea bags

As many tea bags are sealed with plastic and bleached, a lot of people are now choosing to brew loose leaf tea. Canadian scientists found that steeping a plastic tea bag at the optimum brewing temperature releases around 11.6 billion microplastics that we then ingest from our cuppa!

Tea bags can only be put on the compost heap if they don't contain any plastic. A few tea companies have cut out plastic but most haven't to date. If it doesn't say 'plastic free' on the packet then the tea bags aren't! Also, we should avoid the ones in a plastic pouch or plastic-wrapped box if we can.

Alternatives:

- A teapot, and a strainer that goes over the cup.
- A teapot that incorporates a strainer.
- Use a stainless-steel tea-strainer/strainer spoon/infuser with loose tea for a no taint cuppa on the go.

- There are a lot of quirky tea-strainer designs around to suit every taste, and of course it's worth looking around in charity shops. So many strainers/strainer-spoons were put away in cupboards and drawers when tea bags came in. We used to have the strainer-spoon variety in our kitchen drawer.
- If you really want to use a tea bag, then reusable organic unbleached muslin, loose-weave cotton, ones are available online.

Tights

- Avoid Nylon ones if you can.
- They will end in landfill or incineration.

You can buy:

- Bamboo
- Organic Cotton
- Wool for the colder weather

These are all good natural alternatives, and be prepared to pay a lot more, but watch out for mixed materials that contain Polyester.

My dear mum used to cut up tights and use them for plant-ties but that is just putting more plastic into our gardens. We shouldn't do this now that we know more about the dangers of plastic.

Toothbrushes

- Bamboo is the best alternative at present.
- The natural antibacterial and antifungal properties of Bamboo make it the hygienic choice.

- Bamboo is also resistant to natural pests and is therefore farmed organically.
- You can get soft, medium and hard bristles with Bamboo brushes but it is advisable to use a soft or medium one to protect gums.
- Children's ones are available and you can buy sets for family use with different coloured handles.
- Most bristles are sadly still Nylon and may be infused with something else like Charcoal or Castor Oil.
- 'Bite', an American company, and Bambooi.co.uk now offer 100% plastic-free toothbrushes with bristles made of plant material, that is entirely compostable. So, it is well worth swapping to Bamboo to help save the planet from the trillions of plastic ones that end up on world beaches.

You can now get toothbrushes with replaceable heads that just slot in. It remains to be seen how these new toothbrushes with an aluminium forever handle and a replaceable head get on. They still use some plastic, albeit a lot less than conventional ones.

Store Bamboo toothbrushes upright in a cup or toothbrush-holder that will allow the water to drain away from the head that must be kept dry. Don't store used brushes in a cupboard or drawer as they need the air to dry properly. They will go mouldy if stored incorrectly.

Disposal of a Bamboo toothbrush with Nylon bristles:

- Soak the brush in a bowl of hot water overnight and the bristles should come out for you to bin.

- You can pull out the bristles with pliers and then bin them.
- You can saw off the entire head and bin that.
- Bamboo is 100% biodegradable so it shouldn't be a problem in the environment.
- Like the old wooden lolly/popsicle sticks, you can use them as plant labels once the Nylon plastic bristles have been removed. You could collect them and when you have enough, take them to a local allotment or use them as part of a bug hotel. They can also be added to a woodpile for the insects or put in a wood-burner or garden incinerator.
- If you don't have a garden, most local authority household waste sites accept wood waste, which is then made into other products, like chipboard.

Interdental brushes

You can get them made of Bamboo. The bristles are still Nylon but it's better than the whole brush being made of plastic. The Bamboo handle can be disposed of in the same way as a Bamboo toothbrush after the bristles have been removed.

Toothpastes

Toothpaste tubes, in their many forms, are a huge global problem, strewn all over the planet in every country. Not only are the tubes themselves plastic but they contain, among other things, thousands of abrasive plastic microbeads that are used to remove stains. These microbeads are adding to the problem of microplastics going down the drain, continuing on to rivers and oceans to pollute and kill marine life. If one of the ingredients in toothpaste is 'Poly-something', it will be a form of plastic.

Manufacturers need to do more and not just continue down the same destructive path. There are alternatives even to plastic tubes. Sugarcane waste, Bagasse, is one of them that is becoming more commercial now. Some of the smaller cosmetic companies are already using this biodegradable material for packaging their various creams/salves, etc.

Alternatives to toothpaste in tubes:

- Tooth tablets – often come in handy little tins that can be refilled.
- Tooth powder - made of a mixture of Miswak, White Clay, Calcium Carbonate and Mint Crystals.
- Miswak chewing-stick – contains a natural form of Fluoride.
- Natural toothpastes that come in metal tubes as of old. There are even subscription services available to help save money and the planet.

Using tooth tablets:

- Pop one in your mouth.
- Bite down.
- Wet the Bamboo toothbrush and brush as usual to form the foam.

I find natural tablets or paste much better than conventional toothpaste. I also find that my mouth feels fresher and no longer has that nasty 'bottom of a parrot's cage' feeling when I get up in the morning that accompanies chemical/plastic laden toothpastes. I don't suffer from bad breath but even my breath smells sweeter and I no longer have that lingering 'toothpaste taste' all through the day. My teeth look and feel cleaner too.

Tip:
In my experience of using different toothtabs, I would say that it's best to buy them in glass or decant them into a small glass jar. They will melt in hot weather if you buy them in a tin box, or leave them in their often-biodegradable bioplastic packaging. Keep them cool and moisture-free in the bathroom in glass.

Toys in hard plastic

These are a massive global problem when it comes to plastic, and toy companies will have to change their manufacturing methods. A huge source of planetary-pollution is clam shell/blister packets that most toys are packaged in but then most toys themselves are plastic.

The good news is that some toy-companies are working towards more natural materials but the bad is that most still use virgin plastic to make their colourful products and packaging. Bagasse, Sugarcane waste, is now being considered for some toys. Others could be made of Corn or Wheat bioplastic, and, of course, there is Wood. There is also going to be a significant reduction in packaging in general.

Think about our precious children hugging soft toys and potentially breathing in all those plastic microfibres every day.

- Look for toys made of natural materials and fibres.
- Avoid anything artificial, sparkly, shiny or over-fluffy.
- Avoid cheap ones they won't last. These all contain plastic because plastic is the cheapest way to manufacture products.
- Wooden toys are becoming popular again but make sure they come from a responsible and sustainable source. Always check labels first when you can.
- Whirli.com is an innovative subscription toy library that might be worth considering. They are not completely plastic free but you can choose toys that are.

I know it's difficult to persuade family and friends not to give children tons of plastic but be firm, stand up for what you want for their future: A healthy strong body and a safe environment for all life.

If we don't stand up for the world we want right now, nothing will ever change, and Mankind won't have a future!

Travel and holidays

Pack, along with everything else needed:

- Reusable water-bottles, stainless-steel is best
- Reusable cutlery

- Reusable straws
- Bar soaps
- Shampoo bars
- Conditioner bars
- Travel soap - good for hair, body and clothes when travelling.
- A small repurposed lunchbox with compartments in which to store all your soaps instead of splashing out on soap tins, which can be a bit pricey.
- Plastic-free marine/reef-safe sun cream/block
- Cotton/reusable face wipes
- Cotton flannels in a wet bag
- Straw sunhats or cotton hats
- Water-purifying tablets or a special water-purifying bottle that will double up as an ordinary water-bottle, especially if going to Africa or Asia.
- Look for water coolers/fountains in hotels and airports and always ask for refill points.
- Never risk drinking dirty water!
- Don't eat a salad unless you know it has been prepared using clean fresh or boiled water.
- Always buy water if you need to from a reputable company. Do the same with any milk or ice-cream products. Raw milk in hot countries can contain deadly bacteria and can cause Bacillary Dysentery among other things, which, believe me, you don't want to experience!

I nearly died from being poisoned by a contaminated water-supply in India when I was only 24. I ended up with Amoebic Dysentery that developed into Amoebic Hepatitis when it infected my liver. I was seriously ill for months and it took years for me to fully heal.

Ladies:

- Pack a long cool cotton headscarf if you are going to countries that require you to cover your head for religious/cultural reasons.
- Charity/thrift shops usually have a good selection at very reasonable cost.

Ladies and Gentlemen:

- A long cool cotton scarf is one of the most useful items of clothing to pack for travelling wherever you're going in the world.
- It can double as a face-mask in dusty or smelly environments. It also helps to keep hair clean, bug free, and untangled on rough journeys. It is a good idea, and on the whole far safer, for any traveller to blend in with the local people whenever possible.

U

Umbrellas for the rain

Try to buy a good well-made and strong one that will last at least a decade. Some are constructed so the air flows through them and there is no risk of them turning inside out in windy conditions, one of the reasons they break and are discarded everywhere. They are more expensive but we are trying to save on waste. Most are made of polyester and end in landfill worldwide but you can still source umbrellas made of cotton.

Umbrellas/Sunshades for outside space

We must avoid any of the ‘shaggy’ plastic ones. They can shred during storms if left outside and cause environmental damage when in pieces. Avoiding the polyethylene ones which are 100% plastic is also a good move towards a healthier planet.

There are natural ‘shaggy’ ones made of plant fibres. You can also find umbrella-like sunshades made of heavy-duty cotton and Shade sails made of cotton Sail cloth, which is waterproof, hard-wearing and long-lasting.

Underwear

Ladies, and men’s underwear, is difficult to find plastic/ Elastane-free and comes at a significant cost if it is all natural.

The following natural materials are available:

- Organic cotton
- Silk
- Bamboo
- Hemp

Annoyingly, they may be mixed with plastic fibres in some cases, although most artisan/bespoke makers avoid any non-natural materials where possible.

V

Valentine's Day gifts – celebrate Love the eco-friendly way

Love flowers?

- Source them plastic free if you can.
- Some florists wrap them in real Cellophane which can be home composted, paper or Hessian.
- Plant a lovely fragrant rose bush together to give love to Mother Earth too.
- Plant a beautiful flowering tree/fruit tree that will support wildlife.
- Donate a tree to a woodland project or rainforest.
- Give a health-giving houseplant. All houseplants clean the air and add to our feeling of well-being.

Love cards?

- Send sustainable/recycled paper cards and avoid anything shiny or glittery that is plastic.

Love chocolate?

- Source palm oil free with plastic free packaging. Palm oil is causing widescale deforestation and the destruction of Orangutan-habitats.
- Most artisan chocolate comes in traditional paper and foil or in a smart cardboard box.
- Some UK supermarkets stock their own brands in foil and paper.

Love perfume?

- There are lots of chemical free and natural perfumes available.

Love to cook?

- Cook a romantic meal at home for your best beloved or prepare one together.

Love to bake?

- Make a wonderful cake that represents your shared love or create one together. Maybe you could also include a quirky cake-topper for a bit of nostalgic fun. There are great recipes and cake-designs online and in specialist books.

Love wine?

- Choose local organic wine that supports the community and hasn't been produced using pesticides.

Other gifts

- Avoid stuffed toys/hearts/fluffy things, etc. that are contributing so much to plastic pollution worldwide.
- Please gentlemen avoid the lacy lingerie for your sweetheart unless it's made of natural materials.
- Most ladies' underwear is made of plastic/ Elastane now and will not be appreciated if your sweetheart is trying to go plastic free.

- I'm sure a beautiful soft organic cotton or silk nightie would be a wonderful and charming gift.
- Please ladies no plastic/Elastane novelty underpants for your fellas! Natural undies are also available for the gentleman in your life.

Have a lovely planet-friendly day full of love!

Vegetable brush

Look for:

- Wood and plant fibre like Beech wood and Tampico.
- A nail brush made of Bamboo, with natural bristles, will do the job just as well and might be cheaper anyway. (I have found that some natural products have overinflated prices that do nothing to encourage us to go plastic free! I always shop around for cheaper more effective versions.)

W

Wallets

If you want to avoid the plastic or leather type, they can be made of leaf-leather, literally made of leaves from certain trees, or cork from the Cork Oak that is sustainably grown. The beautiful soft bark is harvested with no harm to the tree that will grow a new one. It can be dyed different colours using natural dyes. Some companies even offer engraving/initials on their Cork wallets.

Washing/cleaning the car

- Avoid the washing liquid in plastic if you can and using a lot of precious water from a hose.
- Warm water in a bucket and an old cotton rag/t-shirt and an old towel for drying do a good job.
- Brush the dirt off wheels with a natural hand brush. Use what you have before buying anything else to do any job.
- Look for a natural car-wax in a tin to apply and then buff with soft cotton rags.

Washing clothes in a washing machine

Use Earth-friendly products when you can that are not in plastic.

- Tru Earth Eco-laundry detergent strips are impregnated with natural cleaning ingredients that won't harm the environment and completely dissolve during the wash.
- They come in a paper packet and have all the frees we could desire in an Earth-friendly product.

This is what they say on their website:

'Tru Earth Eco-Strips laundry detergent have a dramatically smaller carbon footprint than liquid and powder detergents. A load's worth of today's leading detergent weighs in at over 40 grams. A single Tru Earth Eco-Strip weighs less than 3 grams, or a whopping 94% lighter!'

Detergent strips are also available in the UK from Simplelivingeco.com.

Soapnuts, from the Sapindus species of shrubs and small trees in the Lychee family that are native to tropical regions of the world, are becoming a popular soapy alternative now. They are available from zero waste/plastic free businesses, and some people swear by them.

We should always put on a full load, especially if some of the clothes are made of Polyester fibre. There won't be as much friction in the drum and the clothes won't shed so many plastic microfibres into the environment via the sewers. There are fibre-catching devices available but the best thing to do is to avoid buying clothes that are made of plastic materials whenever we can. Buying clothes made with natural materials that will eventually biodegrade without harm in the environment is the answer to our over-use of Polyester clothing.

I buy most of my clothes from charity/thrift shops and always go for natural materials. I still have some old 'artificial' ones but I restrict the number of times I wash them. I use the Eyeball and Sniff Test, and only put them on a full, cool short wash with non-bio powder from a cardboard box. These powders are still available from a lot of UK supermarkets on the lower shelves. As I've worked for a couple of supermarkets myself, I know it's well worth looking on the lower shelves as the eye-level ones are there to showcase expensive big brand products, most of which we don't need or can buy a lot cheaper on the lower shelves!

Washing-line

Most is now plastic with a steel core but you can just buy some 100% cotton rope and use that instead. Either tie it to poles or you can buy ring-fixings.

Washing up/dish washing

Plastic sponges/scrubbers harbour bacteria even when washed because plastic is a breeding ground for all sorts of nasties in a wet environment. Also, tiny plastic microfibres released by the scrubbing-action end up in our waterways. To make things far worse, pesticides, dyes and bleaches are used to make these! Do we really want to use them on our family dishes?

There are so many alternatives now to the plastic scrubber/sponge, those terrible plastic-fibre cloths and those cheap Nylon brushes that are frequently found dumped on beaches worldwide.

Options:

- Hands - if you are only using hot water and a mild eco liquid for a quick wash of a few items, or no liquid at all.
- Cotton flannel or t-shirt rag that can be thrown into the weekly wash.
- Cotton string dish mop with a wooden handle – used universally before plastic brushes came in.
- Coconut fibre pot scrubber for those stubborn bits.
- Coconut fibre brush with a long wooden handle. Especially good for getting to the bottom of

mugs. You can get a thinner one especially for bottles.

- Stainless-steel or copper wool scrubbers for stubborn stains but these do eventually break into pieces that go down the drain.
- A stainless-steel Chainmail scrubber that cleans well without a lot of elbow-grease, and will last far longer, is a better option. It is also super hygienic because it doesn't hang onto bacteria the way other materials do. Just rinse after use and hang up to dry. It will clean most pots and dishes but may not be suitable for pans with a ceramic surface. It is especially good for glass and stainless-steel items.
- Loofah, made of a plant. Use slices for convenience. You will need a sharp knife to cut them up. Keep loofahs dry after use as they will go mouldy if left wet.
- Bamboo pot and plate-scraper.
- String scrubber that can be thrown into the weekly wash.
- Knitted string dish cloth that can be thrown into the weekly wash too.
- Wood and plant fibre long-handled dish brush – there are small ones without the handle available and some which include copper bristles for easy pot-scrubbing.

Washing-up bowl

If you need a new bowl and don't want to use plastic that always seems to go manky or breaks, check catering sites for stainless-steel bowls that come in different sizes.

Although they are usually mixing/salad bowls, some are big enough to use in the sink. Just check dimensions first. A 34cm should fit most sinks and with care should be long-lasting. They are lightweight, economical, easy to clean and sanitise. They also keep water hotter for longer. You can still buy enamel bowls but they might be a bit large for most modern sinks as they were used in the larger Butler sinks or on tables in the past.

Square stainless-steel washing up bowls are available, but far dearer as they are meant to last a lifetime. Of course, Stainless-steel can be recycled with the metal recycling if it ever needs to be.

Normann of Copenhagen do a rubber washing-up bowl. It's also expensive compared to plastic. As it's made of a natural material, it's biodegradable at the end of a useful life.

Washing-up/dish washing liquid

I've gone no washing up liquid. The most important ingredient in any washing up is very hot water.

Washing-up/dish washing bars are an expensive luxury and according to some don't do a good job. I find that a bit of ordinary non-scented soap wiped around the object and a good hot rinse works better. As long as dishes are squeaky clean, you're fine. Just dedicate a bar to washing up only, and try to source it plastic and palm oil free if you can.

Cleaning icky pots, etc.:
I just give them a good scrub after a soak using a coconut or stainless-steel chainmail scrubber, and a little

ordinary non-scented soap, finishing with a good hot rinse. Some people use Bicarbonate of Soda or Lemon for greasy ones. I haven't found it necessary as I'm a vegetarian and also don't use a lot of fat or oil.

Yes, the soap can make a mess in the sink, especially a stainless-steel one, but I remind myself I'm doing it for the planet and just roll up my sleeves and get on with the cleaning.

It is a bit odd after nearly a lifetime of putting a squirt of liquid in the water but, honestly, I feel healthier for not using detergents. Of course, it is a personal choice and there are eco alternatives to the petrochemical ones. By the way, all petrochemical washing up liquids have a pesticide in them, called an insecticide but it is still the same thing, toxic. Do we really want our families ingesting that?

Try to avoid products that contain palm oil whenever possible if they don't have a sustainability label on them. Palm oil cultivation is destroying whole ecosystems in Indonesia, with slash-and-burn techniques, especially in the forests where the now critically endangered Orangutan lives.

Palm oil is found in hundreds of products, like washing-up liquid, shampoo, spreads, peanut butter, chocolate spread and ice cream, and is often sneakily 'hidden' under a couple of dozen different names that include:

- Palm kernel
- Hydrogenated palm glycerides
- Sodium lauryl sulphate
- Stearic acid

Many palm oil producers are small farmers who care for the environment and need the income for their families and communities. Destruction is caused by the huge concerns that clear large swathes of land. Look for Green Palm Sustainability labels on products that contain palm oil. The Rainforest Alliance has a certification as well. You can also find the Palm-oil sustainability label on products that features an Orangutan.

In April 2021, Sri Lanka banned imports of Palm oil. They also told producers to uproot their existing plantations, and to replant native forests, because they had taken over the country with vast monocultures. It will be a phased programme of uprooting, being replaced by Rubber trees and environmentally-friendly crops.

If in doubt about any product, contact the manufacturer for answers.

Waste paper bins/baskets

These are lovely natural additions to any home or office:

- Willow – traditional and will last decades.
- Bamboo – the fastest growing grass on the planet, and long lasting.
- Seagrass - flowering plants which grow in marine environments and make strong baskets and even rugs and flooring.
- Metal mesh - lasts forever when kept dry.

Watches

- Most are made of plastic and have plastic bevels and plastic straps, so avoid those if you can.
- Fully stainless-steel is a long-lasting affordable option.
- You can also buy watches with wooden or Bamboo dials.

Water-bottles

We can stop using plastic ones and save ourselves a fortune by buying a reusable stainless-steel one that will last a lifetime. Small lighter bottles are available for children, with a built-in straw in some cases. Most supermarkets stock the adult-sized ones and the children's can be found in zero waste/plastic free shops and online businesses. They are becoming popular now as parents look to their children's plastic free future.

You can also find:

- Bamboo – buy one that has good credentials.
- Glass – not ideal for the traveller but these are usually the ones that contain the charcoal filters for purifying water.

Water butts for outside spaces

- Most are made of hard durable plastic but I would be more concerned with storing precious water on a drying planet than worrying about whether this particular plastic can be recycled in 30 years' time!

- Wooden water-barrels and aluminium ones are still choices if you want to go completely plastic free but probably won't last as long in the elements.

Watering-cans

Well-made and long-lasting metal/aluminium ones are still available. Buy one with a good watering-rose on it for gentle soaking when plants need it.

Wellingtons/Wellies/Rubber boots

Try to avoid cheap boots with a lot of patterning as they are usually plastic.

Look for:
Adults/Children's Natural 100% Vulcanised Rubber boots with a cotton or wool lining.

Pokeboo make ultralight packable/foldable rainboots in natural rubber. They are comfortable, waterproof and friendly to the environment. These boots can be easily stored inside their carrying case, which comes with a carabiner for easy attachment to backpacks, etc. Pokeboo won the Good Design award for their ingenuity in 2018.

You can turn old boots and shoes, preferably in natural materials, into quirky planters, and even birdhouses, for the garden or outside space. Make a few holes in the bottom for drainage if they aren't already 'holey'. Fill the planters with compost and plant something pretty and trailing. Nail/fix the birdhouses high up in a tree for safety, where it is warm but not too warm for the eggs and finally the chicks. Sit back and enjoy your new neighbours!

Window/glass cleaning

- Regular White Vinegar or Apple Cider Vinegar in a glass spray bottle or reused plastic one is a good alternative to chemicals in plastic. Use with a cotton rag that can be washed.
- Scrunched up newspaper has been used for decades. It is effective on windows and mirrors, and for cleaning greenhouse-glass, to finish off and give a smear-free clean.

Windscreen scraper/cleaner

- Avoid the plastic ones if you can.
- A wooden or Bamboo kitchen spatula makes a good strong scraper for snow and ice.

Wipes

Wet wipes, or Moist Towelettes as they were first known, were invented in the USA in 1958 by an enterprising man called Arthur Julius who then struck a deal with Colonel Sanders in 1962 to provide a free towelette with all Kentucky Fried Chicken meals.

I certainly don't remember them here in the UK until the 1970s. What I can remember clearly is that they had a strong and sickening chemical smell the second they were released with a flourish from their tiny, sadly, treasured plastic packet. I've had some nightmare journeys in the past because they made me feel so sick on top of being a terrible traveller.

Wet wipes were designed to be tough, made of Polyester, and disposable and were mostly used for cleansing sticky hands. (My dear Mum always used a flannel in a toilet-bag.) Now there is a wipe for everything you can think of and a few you never thought of. The infamous 'fatbergs' of the sewer systems are caused by hundreds of thousands of these squares of plastic fibre being flushed down the nation's toilets, because they were wrongly labelled 'disposable'. They mix with fat, oil, grease and only God knows what else. A truly vile manmade horror for the sewer-workers to constantly clear!

Oceans start in the toilet. No wipe should ever be flushed!

Modern people should learn to wash with a flannel and hot water and not just wipe with an unnecessary antibacterial chemical that may even be compromising our immune systems.

Not only are these wet wipes costing the water companies millions of pounds/dollars to clear each year but they are costing the Earth, that shouldn't have to endure plastic pollution of any description! There are now so-called 'flushable' wet wipes but they are still causing problems and should only ever be binned to be on the safe side.

We must never use kitchen roll as a substitute for toilet paper in times of shortage. It was never designed to go

down the toilet. It's made to soak up liquids like a sponge and will therefore clog sewers and add to fatbergs. Remember:

Only
Pee
Paper
Poo
Goes
Down the Loo!

Definitely not disposable plastic contact lenses, tampon applicators, or cat-litter, which is full of pathogens that will harm aquatic life! Anything that is not designed to go down the toilet shouldn't be flushed away!

Cleaning surfaces

Keep a handy Rag Bag in the kitchen. Hot water or steam with a soft cotton cloth or an old t-shirt rag is just as effective as chemical-laden sprays. You can clean toxic free at no cost to you or the environment.

You can also use cut up:

- Cotton knickers and pants
- Cotton nighties and pyjamas
- Old terry cotton bathrobes or towels
- Flannelette sheets

Bicarbonate of Soda, Lemon or Vinegar can be used for tougher cleaning.

Woolly hats/Beanies

- Most are made of Acrylic or Polyester fibre which is not healthy to breathe in.

- The thermal ones are plastic too and should be avoided.
- Wool ones are becoming popular again because they are so lovely and cosy and stay on the head without slipping around! They are more expensive but will last a long time if they are cared for properly and according to the washing instructions.
- Knit your own if you can.
- There are some wonderful patterns online.

X - I can't think of anything for 'X' apart from the X-box, which has no alternative, x-ray, or xylophone which is metal and wood anyway!

Y

Yogurt

So much of this is in plastic. Some companies have started to use glass but usually with an unavoidable plastic lid.

In 2020, one popular supermarket removed all the see-through plastic lids from its own label yogurts. Before I stopped buying them altogether in plastic, I used to leave the lids on the shelves. So, I'm really happy about this tiny but important change that we can only hope is taken up by other supermarkets and dairy producers as soon as possible.

The electric yogurt makers available can involve a lot of plastic and plastic pouches of flavourings. You can get stainless-steel ones that are ice cream makers too.

Yogurt is actually simple to make in a vacuum flask overnight but you will need one pot of live yogurt as a starter. After that you can keep a little of what you make each time as the following starters. You won't need to buy any more yogurt in plastic after that.

You will need:

- A saucepan
- A thermos flask, preferably with a wide neck
- A food thermometer
- A measuring jug if you need one
- 1 pint of whole milk – you can use goat and sheep milk as well
- 1 tablespoon of live yogurt
- Fruit of choice or Vanilla food essence

Method:

- Slowly warm the milk just short of boiling on the stove.
- Cool it to 110 degrees and stir in the live yogurt starter.
- Pour the mixture into the thermos and close securely.
- Leave it overnight or for 6 to 12 hours. There is no need to refrigerate at this 'cooking' stage.
- After that time, stir in any flavouring or fruit you want.

The longer you leave it the more tart it becomes. If you prefer a mild flavour, eat it straightaway. If you like it a bit sour, leave it a few days extra in the fridge.

Z is for Zero waste

This is what we should aim for but until that happens in every country throughout the world, we can only do our best in our own personal circumstances.

Everything we do helps create ripples of change across the globe.

Tiny positive daily actions are important and can be achieved by everyone.

Last thoughts

What a plastic-free world would look like

'Another world is not only possible; she is on her way. On a quiet day, I can hear her breathing.' – Suzanna Arundhati Roy

When the Coronavirus Covid-19 pandemic ends I'm not sure governments will push hard to ban plastic. Worldwide, our health professionals desperately need Personal Protective Equipment as well as other medical necessities, most of which are throwaway plastic. There are alternatives that could be explored for medical use. Let's end our dependence on toxic plastic altogether.

In the USA, there have been reports that small items of PPE are being flushed away, clogging sewers with dangerous hazardous material. Here in the UK and elsewhere plastic masks and gloves are discarded on the streets and in rivers and oceans, without a thought for the serious consequences of pollution and infection.

I can hardly believe the sheer stupidity of these actions! The advice about masks and face-coverings has been confusing for us all. I'm afraid the supermarkets and pharmacies are just adding to single use that is creating a new and dangerous litter-problem.

On the one hand pandemic experts tell us that plastic harbours viruses. On the other, our health services cry out for more. The plastic industry is rubbing its hands with glee with the perfect excuse to exploit the crisis, discredit reusables and block laws prohibiting single-use.

Back came single-use plastic bags with a vengeance at the start. This was disheartening after we had fought so hard to get them banned. Published studies from virologists show that this particular Coronavirus can persist on plastic surfaces for up to four days. This is the longest duration among all materials tested and one of the reasons A Plastic Planet launched the world's first plastic-free PPE in 2020.

In the Philippines, the Abaca, a relative of the Banana palm, is now being used to make masks that biodegrade after a few months. This is a great example of using a natural resource. The fibres could also be used for hospital gowns.

During this pandemic it is obvious that the great work we Eco Warriors have done beforehand is falling on deaf ears. I personally find it difficult and frustrating when I am forced once again to buy food in plastic because there is no choice in my town. Price is one of the biggest obstacles to going plastic free, made even more difficult during Lockdowns when stock is low. Thank God my Council still recycles and incinerates waste while some have put a halt to it. I sincerely hope we learn many valuable lessons about how to treat people, animals, and the environment during this heart-breaking crisis.

One of the results of the easing of Lockdown in June 2020 in the UK was the devastation 'visitors' caused to wildlife reserves. Littering, lighting fires in wild places, destroying nests and wildlife became major concerns for the dedicated staff. This was the result of sheer carelessness and the boredom of being cooped up for so long. I doubt these people had ever been to a nature reserve, and therefore didn't know how to behave in a respectful manner. As far as they were concerned, they were 'free' to do what they liked. (Those blasted single use BBQs need to be banned worldwide for the fire and litter hazards they are, and for the stupid irresponsible behaviour they encourage!)

We need governments that truly care and are not afraid to be firm when necessary. Curfews, closures and bans should be used when there is danger to life.

We'll end this plastic nightmare together!

Billions of people can force sweeping changes. Governments have completely forgotten they serve **us**! I want to look back in years to come and see what has changed. I'm delighted to say that I can hardly keep up with the changes happening worldwide. (At this point I must apologise if I have mentioned a ban that didn't come to fruition because of the pandemic. Some governments may have backtracked on their commitments.)

At last, we are waking up to our trashed planet. Everything is vitally connected and should be kept in balance. We must be like Ants, diligently working for the common good.

The United Nations is adopting a new framework that will include Nature when measuring economic growth and well-being. Gross National Happiness is a philosophy that guides the government of Bhutan already. Nature plays a big part in the life of this beautiful country at the top of the world.

We have a place on our fabulous planet as loving Guardians, not Possessors.

Do bans work?

They do but if governments don't specify what can or cannot be used, confusion reigns and people will push any boundary. Banning the production of plastic worldwide is the **only definitive** answer to the plastic crisis coupled with a major planetwide clean-up.

It's easier to implement new systems in small countries of only a few million citizens. Scaling up has its problems. It always takes time as well as a great deal of money, but we must do it. Good things are starting to happen all over the globe.

Africa is the world leader at present, with bans being implemented in 34 out of 54 countries. Sixteen have totally banned or partially banned plastic bags. As early as 2008, Rwanda banned non-biodegradable plastic bags. This prohibited their manufacture, use, importation and sale. Visitors to Rwanda are not permitted to take this type of carrier bag into the country. Rwanda is one of the cleanest nations in the world because of its Earth-friendly policies, reforestation programmes, and monthly litter-picks.

The rest of the world urgently needs to take a leaf from their book! We must insist on safe natural alternatives to replace all single use. Fortunately, there are many already in existence in Africa and where traditional crafts are still employed.

In 2019, Oahu in Hawaii signed into law one of the nation's strictest bans on plastic. It prohibits single-use plastic plates, bowls, cups and general service-ware by 2022. Naturally, there was opposition from the food industry but fortunately the majority ruled in this case.

There was more good news in 2020 when China started to phase in bans on plastic bags and other single use plastic. This will be another huge blow to the industry from this vast country of 1.44 billion that uses so much in everyday life. The hardest task will be to get the whole of the USA onboard where most of the material is produced. Some States implemented their own bans because the administration seemed to avoid discussing the subject of saving the planet altogether. In 2020, New York State banned plastic bags and the Washington Senate passed a ban on the sale and distribution of most expanded Styrofoam products.

Some single use plastic products were also banned in France in January 2020, which has the goal of phasing out all single use plastic by 2040. That's a full 5 years ahead of the UK.

Costa Rica banned Styrofoam in 2020 and drastically reduced all single use plastic. From now on, no petrochemical plastic will be permitted and must be replaced by alternatives that are 100% recyclable or biodegradable.

There was more good news at the beginning of 2020. Traders in Tamil Nadu, South India, banned the sale of some branded drinks in favour of local ones and Coconut water. The top two associations of traders that proposed the ban say that soft drinks companies take far too much water from rivers. This leaves farmers struggling to irrigate, especially in times of severe drought. Of course, it will reduce the amount of plastic too.

At the end of 2020, Indian Railways' Chai-sellers went back to using clay cups for their tea instead of plastic. The cups, that are smashed on the tracks, return harmlessly to the earth, and there is continuous work for the potters.

Most everyday plastic products will be banned by the Central Indian Government from the beginning of 2022. It's a well thought out and comprehensive list covering manufacture, import, distribution, stocking and sale. I just hope it still goes ahead.

All these bans are bound to be hugely effective and will definitely impact the plastics industry. I can't wait to see what the knock-on effect will be throughout the world!

'The rules of our world are laws, and they can be changed. Laws can restrict or they can enable. What matters is what they serve. Many of the laws in our world serve property - they are based on ownership. But imagine a law that has a higher moral authority... a law that puts people and planet first. Imagine a law that starts from first do no harm, that stops this dangerous game and takes us to a place of safety...' - The late Polly Higgins, campaigning lawyer for the introduction of the world crime of Ecocide - 2015

Ecocide is a crime against Nature and Life itself, and a reason for the impeachment of presidents! Let's grant the planet Personhood and make Ecocide a world crime. France is starting the ball rolling by making Ecocide an 'offence' punishable with heavy fines and, depending on the severity of the offence, time in jail. This new law needs real teeth and it appears it's getting them. Thank you, France! Now the rest of the world MUST follow!

Plastic has simplified our lives to the detriment of our glorious planet

Now systems and infrastructures have to become complicated. The solution to everything can no longer be PLASTIC and we must build a new Earth using sustainable materials that embrace traditional skills.

We can't fall back into the trap of vast monocultures that deplete natural resources. One of these is the huge Agave/Sisal string plantations on the richly biodiverse island of Madagascar, that are impacting Lemurs. A reduction in genetic diversity means that crops have a lower resistance to the diseases that now plague them worldwide. We need to restore balance and also safeguard animal habitats for the future.

Across the globe, our young people are taking action. When I go out with my Alternatives to Plastic Demo Table it is the young people who are already onboard. I sometimes can't get a word in edgewise. Our simple Ocean Fishing Game, see **Games to raise awareness,** always draws a crowd and it's easy to make something similar to add to any Eco table for a bit of fun education.

We have a sacred duty to talk to the younger generations about loving Mother Earth. I think we ought to teach our children as soon as possible how to co-exist with Her. There are wonderful books available on this subject as well as children's Eco magazines. All schools should have at best a wildlife garden and at least some sort of growing space to connect them to the earth.

Pristine beaches should be left exactly as we find them. If they are a bit dirty, we can pick up some of the rubbish and dispose of it mindfully. No stones or shells should be taken as they are a vital part of any marine environment. Shells are used by Hermit crabs as protective homes for their soft bodies. They need larger ones as they grow. Sadly, according to recent studies around the world, they have been using plastic caps and small pots which are totally unsuitable. Once they get in, they can't get out, and soon die.

Natural resources have been used for centuries

I remember using Banana-leaves as plates in India. Coconut-shell was used for bowls and other crockery and even made into spoons. Newspaper was used to wrap everything and made into carrier bags as well. Brooms and brushes were constructed of soft twigs. Any waste natural resource should be used whenever possible and will add to the local economy.

When plastic is no longer an option, creativity blossoms!

Manufacturers should not be using their unique plastic-free inventions to line their own pockets! They must share them with the world. They must also take responsibility for the end of life of all their products. They can't be allowed to flood the world markets with things that will never biodegrade.

The devious practice of planned obsolescence must be outlawed worldwide to address dead-tech and white goods mountains. Let's bring back pride in a job well done. Manufacturers should be given financial incentives to make replaceable parts for future tech. They must recycle/repurpose whatever they can.

Industry will be forced to change.

The European Commission stated at the beginning of 2020 that it would introduce new waste-reduction targets, coupled with sustainability laws, to ensure that products on the EU market are not only recyclable, repairable but designed to last. Their target is to halve waste by 2030.

Cheap shoddy goods have been the downfall of our societies and the destroyers of the planet! We must buy to last.

Natural alternatives **are** cautiously being developed. Brown Seaweed has been made into Water-blobs and has already been used successfully for marathons. The exciting thing about this material is that it could replace many of the petrochemical plastic films and pouches used in the food industry.

Biodegradable glue is being used to stick cans together for multipacks. Edible yeast rings that can be eaten by wildlife and sealife are being made instead of the tough plastic ones. (Any plastic rings, that often trap wildlife, should be cut up before disposal.)

Creating bioplastics from waste farm produce is another way forward. Many petrochemical plastic films and boxes/food packaging can be replaced by such materials that will not harm the environment or us. Even Milk can be used.

What else is being developed to replace plastic?

- Orange-peel is being used to produce silk.
- Bio-engineered spider-yeast is also being explored by the fashion industry.
- Corn and Wheat bioplastics can be used for many things.
- Faux fur is also being developed using corn.
- Coffee-grounds can make a cotton alternative, and even coffee cups.
- Avocado seeds can be turned into cutlery.
- Edible wheat-waste coffee cups have been invented.
- Compressed Banana-leaves are being used to make plates.

- Coffee husks are being used in Colombia, where most of the world's coffee comes from, to make building materials for low-income housing.
- Fruit leathers are becoming popular with clothes-designers.
- Even mushrooms are being used to make a plant leather, called Muskin, for bags and hats. Maybe the mushroom-leather could also be used for pet collars and leads.

I would love to see these 'Leathers' take off. It would be wonderful if they could completely replace animal-leathers for items such as shoes, jackets, belts and bags. Apart from the unnecessary deaths of sentient beings, tanning leathers requires the use of powerful toxic chemicals that seriously impact the environment.

Alternatives to plastic should be employed in every industry but I do think wood has a big part to play in future. I love trees but if we use more wood, we will need more trees. It will become a regenerative cycle. Trees are now known to increase local rainfall coupled with cooling the planet. Buildings constructed of wood sequester carbon. They also have huge health benefits compared to brick and concrete. Using wood in building-construction also avoids the need for plastic paints and wallpapers.

Utilising reclaimed timber and storm-damaged trees is also what we need to do. Nothing should ever go to waste.

One of the unlikely raw materials to make new textiles is cow-dung, which can also be used to produce energy.

Dutch designer Jalila Essaïdi proposed that the clothing industry should use a bioplastic made from regenerated Cellulose fibre from the dung. A method known as Mestic. This could replace Cotton, a water-greedy plant which is one of the crops that has a high carbon footprint. It takes 400 gallons of water to produce one Cotton tee-shirt. This is not sustainable for a future where water will be scarcer on a warming planet.

Some airlines have gone plastic free. They have even handed out stainless-steel water-bottles. Airlines should go back to serving food on ceramic plates, with metal cutlery and utensils. You can still find this in first class so it can be done throughout. Why some have gone down the pointless biodegradable route is beyond me. Airlines should employ dishwashers!

The world fishing industry needs an urgent overhaul and there should be eye-watering fines for dumping rubbish and fishing-gear from ships plying world seas. Even the confiscation of vessels needs to be a threat to serial polluters. The world also desperately needs domestic and commercial biodegradable fishing-gear so anything lost will not harm the environment.

Our whole world needs to slow down!

If eating and drinking while walking and on public transport were to be banned then people would have to take the time to sit down and eat and drink properly. Workers would need a full hour for lunch in order to do this. It would make an enormous difference to food-litter problems.

We can't stand idly by and do nothing when we see pollution. We must push world governments and authorities to implement clean-ups and educate societies not to foul their own backyards! Good ideas must be given voice, implemented as soon as possible, and spread around the globe.

We need real money spent on practical and mindful solutions.

According to a review, published by a group of prominent environmental scientists, on the 1st of April 2020 by the journal Nature:

'Recovery rates across studies suggest that substantial recovery of the abundance, structure and function of marine life could be achieved by 2050, if major pressures — including climate change — are mitigated. Rebuilding marine life represents a doable Grand Challenge for humanity, an ethical obligation and a smart economic objective to achieve a sustainable future.'

We need to buy for life again. White goods must be manufactured to last a lifetime. They are an expensive buy for anyone, and we should not be forced to constantly fill the coffers of unscrupulous producers who only care about profit!

We can't let Earth suffocate under our rubbish!

We have to re-educate the public about everyday plastic use. Many of us still buy the thicker plastic bags for life. I often see them put out with rubbish in them instead of being reused. If this isn't discouraged or outlawed worldwide, these bags will continue to be a scourge on the planet. One major UK supermarket stopped supplying them in 2021. I hope this will encourage others to do the same.

Time is running out and there is no quick fix!

We have to cut out plastic wherever it's hiding. I'm so sick of seeing waste lying around with little or no attempt to clear it up. Every pair of hands is needed and it's good to see people all over the world coming up with new ways to clear plastics from the environment.

One of the most effective ways to clean up marinas is to use the clever Seabin, invented by two Australian surfers, Andrew Turton and Pete Ceglinski, that gently sucks in floating rubbish but doesn't harm the wildlife.

Fionn Ferreira, an Irish engineering student, has developed an award-winning way to magnetize plastic so it can be removed from water. While Hawaiian engineering students have come up with a vacuum-cleaner that sucks up beach plastic, separating it from the sand and returning it clean to the beach.

Verdi Gras

There's a movement in the USA to green Mardi Gras, the yearly festival for Shrove Tuesday, which has become a serious plastic problem in New Orleans. Each year between January and the actual day, 75 parades make their way through the city. This celebration generates a great deal of money for the local economy. The Verdi

Gras organisation aims to make this colourful festival more sustainable by encouraging the sale of biodegradable paper beads instead of plastic ones that litter the streets after the fun of 'The throw'.

Frighteningly, e-waste has been discovered in some of these plastic beads.

Where there is a will, there is a way. There is such an exciting world on the horizon!

Please don't give up pushing forward with this global war. It is a war and there are casualties. Many environmentalists and Climate Warriors around the world are being openly murdered, arrested or just disappear. Powerful greedy people and criminals want them silenced. Don't forget there was huge opposition to slavery and suffrage but People Power won. In time we will win too!

The late American folk singer and social activist, Pete Seeger, could see what we need to do now in his song 'If it can't be reduced'.

'If it can't be reduced, reused, repaired
Rebuilt, refurbished, refinished, resold
Recycled or composted
Then it should be restricted, redesigned
Or removed from production'

We have to build a greener future

All life, great and small, is vital to the health of Earth. We need to know what is going on with the future of plastic. Of course, a lot of us are hoping it doesn't have one. It seems that one solution is to turn it back into fuel-oil. Another is to make every type recyclable. Neither will help us or Earth.

- We must avoid buying as much hard and soft plastic as we can.
- Getting the Petrochemical Plastic Tap turned off will start the healing process.
- Sustainable bioplastics that do no harm to Earth must be developed.
- Alternatives to plastic must be cheaper/government subsidised or nothing will ever change.
- 'Hands off' policies do work and Nature can recover.
- Profit should not come into it when taking pollution out of the water or off the land. We must do it for Earth.
- If we don't vote for people who care for the planet, there will be no planet to care for.

We must all get off our bums/butts and do something!

Buy a litter-picking kit and get on with it!

'Sewage surfer'

In 2016, Justin Hofman of Sea Legacy was taking pictures off the coast of Sumbawa Island, Indonesia when he noticed a tiny Seahorse floating by with its tail curled around some seagrass. As a strong current hit, it released the grass and attached itself to a pink plastic cotton bud. This prescient image went viral and has become a powerful symbol of the world movement against plastic. He said there was far worse pollution in the sea around him but this was the image people would respond to. He subsequently won Photographer of the Year for his 'Sewage surfer', a shameful indictment of our filthy Planet Plastic!

Is this what we should be doing with waste?!

I have mixed feelings about using closed landfill sites as nature reserves. Plastic and other toxic waste like asbestos, added to anything organic, is still there gurgling away underneath.

Building housing estates on old landfill sites to me is even more insane! The natural Methane gas can be extracted along a network of pipes embedded in the sites but I find this disturbing. Surely, it's all in danger of collapse at some point. Also, the waste is entombed in black plastic before the site is backfilled with soil and closed, further adding to the toxic plastic problem. I gathered this information from someone who had once worked at a landfill site, and was absolutely sickened by what he disclosed.

We urgently need to reduce waste globally!

Regift!
Sell!
Donate!

Most returned goods go to landfill or are incinerated because it's cheaper for companies to dump unwanted items rather than put them back into stock.

We need passionate and caring people to lead the way!

Facebook has many great Eco pages. I'd like to say a big thank you to everyone who is fighting for our beloved Earth.

My own page on Facebook is: facebook.com/PlasticsBeGone. I would also like to thank my followers. You inspire me and give me hope.

We must make our voices loud!

- Kick the plastic habit and be a fine example for others to follow.
- Vote with your wallet.
- We don't need to buy what they're selling.
- Pester!
- Complain!
- Campaign!
- Sign and share petitions.
- Ask companies for plastic free deliveries.
- Bring back the Milkman!
- Share the horrors of pollution and the solutions on Facebook and other social media.
- Change banks if they support fossil fuel.
- Message/email/tweet MPs/politicians when you can to encourage them to always stand up for Earth.
- If we don't vote for people who care for the planet, there will be no planet to care for.
- Vote for green policies!
- We must be the seed-planters of change.
- Don't wait for change. Make it yourself!

- Yes, we can start to clean up the planet!
- Buy a litterpicker, take some rubbish-bags, and get on with it!

Remember that plastic is in the dust around us, that we are breathing in, so we must avoid any 'fluffy' clothes or fleeces or using microfibre cloths, or wipes that are 84% plastic. We should always choose natural clothing, furniture and home accessories, bedding and even toys if we want our homes to be plastic-dust free.

In the UK 2020 Budget, a Plastic Packaging Tax was applied at the rate of £200 per tonne of plastic packaging which did not contain at least 30% recycled plastic. This applied to plastic packaging which had been manufactured in, or imported into, the UK. This tax was supposed to slow down the use of virgin plastic but it's not enough to even put a dent in its overuse. As I have said throughout this book, recycling plastic doesn't make a jot of difference. It just puts a rubber stamp on more production.

If we love the planet, we can't love petrochemical plastic!

If we are not part of the solution then we are definitely part of the problem. The old way is throwing it out. The new way is finding ways to produce it without petrochemicals. Then using it for as long as possible before ultimately recycling/repurposing it. We all have our part to play. Are you ready?

It's entirely our choice for the future: Healthy and thriving planet for all life or polluted plastic nightmare of global proportions!

'When I was a boy and I would see scary things in the news my mother would say to me, 'Look for the helpers. You will always find people who are helping...'
To this day especially in times of disaster, I remember my mother's words, and I am always comforted by realizing that there are still so many helpers - so many caring people in this world.' - Fred Rogers

Yes, let's remember, especially in troubling times, that kind caring people are everywhere and willing to lend a hand when we ask.

We need action now! If our global governments aren't going to make the right sort of laws in time then we must push forward and do what we can today!

Keeping informed and ahead of the game is the way to beat this planet-wide crisis. We must also keep telling ourselves that we can do it. We need Green Governments worldwide coupled with a spiritual and moral shift if we are going to save ourselves and Earth from an unbearable future full of manmade horrors.

With all those wonderful 'helpers', here's to the end of Planet Plastic and the return of an abundantly-diverse Planet Earth!

Games to raise awareness

These are just a guideline as I'm sure there are lots of clever people among you who can think of entertaining ways to educate your own children on caring for Earth.

Ocean Fishing Game

You will need:

- Small plastic items that can be hooked out. These can be sourced from games in charity/thrift shops or you could use broken toys your children no longer play with. Stick on a paperclip if it could be hard to fish them out.
- A large shallow cardboard box – supermarkets have these.
- Blue tissue-paper
- Real Cellophane to keep everything clean and give a watery effect – ask a florist.
- An old flattened out cereal-box on which to draw the creatures.
- A drawing-pencil
- An eraser
- A good sharpener
- Coloured pencils or watercolour pencils/paints for the sea-creatures.
- Scissors

- Some planet-friendly glue or paper tape to stick on the sea-creatures. Just do a DIY double-sided roll on the back with any tape and press down firmly.

Fishing rods:

- Bamboo skewers
- Cotton string
- Large recycled box staples, open, for the hooks. You could open out a couple of metal paperclips for the hooks.

Method:

- Cover the inside of the box with the tissue paper and stick it down with a small amount of tape on the sides.
- Draw, colour and cut out your lovely sea-creatures. Have fun with them! Then arrange and stick them down securely on the bottom with some tape on the back. When all the rubbish is gleefully fished out, the children are often surprised by the wonderful creatures revealed, so draw some of the more unusual ones for them to learn about.
- Over the top put the Cellophane and stick that down with a small amount of clear tape.
- Tie the string to the skewers for the rods and attach the box staples/paperclips as the hooks.
- Done!
- Get fishing!

It's a fun game and powerful practical message they should remember forever.

Rubbish/Trash/Beach Clean Bingo

This is another fun activity for all that raises awareness of the horrendous problem world beaches are now facing. You can download cards from online sites or make your own. If you have an artistic flair, why not give each item a personality and a face: Mr. Plastic Bottle, Mrs. Flip Flop, etc. We all love a good laugh and children respond better to humour.

Suggested images:

- Plastic bottle
- Crisp packet
- Cigarette packet
- Cigarette butt
- Sweet wrapper
- Flip flop
- Plastic straw
- Plastic toothbrush
- Plastic comb
- Plastic toy
- Cotton bud
- Plastic washing up brush
- Plastic food-box
- Fishing line and hook
- Barbeque grill
- Piece of plastic rope
- Plastic fishing net
- Tin can
- Glass bottle

One of the most common items is a full plastic nappy/diaper but I don't think anyone would appreciate having to note that down!

Recycling game

Can be adapted for any street clean too. A commercial one is available but why not make your own. Make cards from old cereal packets with various items drawn on them to be recycled in the correct bins. Have fun and include some quirky and amusing things.

You will need 5 pots or containers.
Bins labelled:

- Paper and card
- Tins and metal
- Plastic – not keen on this, as you know by now, but recycling is the only option for some who have no other way of disposing of it.
- Food/organic waste
- Garden waste

Setting up an Alternatives to Plastic Demo Table

Thank you, once again, for being one of the beautiful caring souls that is changing the world for the better. This particular project is to raise awareness of the local litter problems that now blight every community on Earth.

Step 1

- Spend an hour walking around your town/village/beach and really take notice of the crisis. Take a litterpicker and a big bag with you too!
- Collect as many plastic bits of rubbish/garbage you can that demonstrate the local and global problem.
- Ask family and friends to pitch in.
- Put it in a sealed see-through tub or bin, because you want the plastic and dirt to show!
- Fill up a large sweet/candy-jar if you can get one.
- Even a large plastic drinks bottles will do nicely. Glass ones just aren't big enough.

You have to shock people's socks off! They need to be aware of how much plastic there actually is strewn around our precious planet, and especially in our wonderful oceans. There is no point in **only showing the alternatives** because people just won't register. Don't blind people with science or statistics or you will lose them! Make it really simple and visual for everyone, and fun, especially for children and young people. Be as quirky and creative as you can!

The filthy plastic I use on my table was picked up in a couple of hours in one small area of my town. I replace it whenever I have to. I put all the 'plastic baddies' on one side of the table and all the 'goodie alternatives' on the other, with an Atlas showing the 5 Gyres in the middle. Check out the 5 Gyres organisation for more information.

Step 2

- Gather together alternatives. Online searches will help. Remember to use Ecosia and plant some trees while you search.
- Ask family and friends what they have that you can use. You might be surprised to find Eco Anti Plastic Warriors among them already!

You will need these on your table:

- A stainless-steel water bottle and a plastic water one to show the problem.
- A good plastic reusable replacement – all supermarkets sell these now. It's not perfect but it is a good cheap starter for individuals and families to kick the habit. Not everyone can afford stainless-steel bottles although our budget supermarkets sell very affordable ones now.

You will also need:

- Reusable cotton or string bags. Make it loud: **Plastic carrier bags are a scourge on the planet!**
- A keepcup made of natural materials and a typical plastic-lined takeaway cup.
- A Bamboo toothbrush with plant-based bristles and a bright plastic one. Toothbrushes turn up on beaches worldwide.
- A wet wipe and a cotton flannel.
- At least 1 plastic-handled cotton bud and a few Bamboo or paper ones if you live outside the UK. (Plastic ones are banned in the UK but often come in plastic boxes, although they can be found in cardboard in plastic free shops and online businesses.)

Wet wipes should never be put down the loo/toilet! They clog sewers and eventually end up in waterways, seas and oceans. So have a demo packet of those to hand. They are another scourge on the planet at present!

This is one of the important signs I have on my table:

Only
Pee
Paper
Poo
Goes down the loo!

You may want to include some powerful ones of your own if you have the room on your table or on a wall or easel.

Try to source the following extras for your table.

Whatever you can find that is a natural alternative to plastic is a bonus. Be creative and add some quirky items to bring a smile. Laughter is a great educator.

Washing up and cleaning

- A plastic washing up brush and a wooden or coconut one. Plastic bits go down plugholes!
- A coconut fibre dish-scrubber, a plastic green one and a natural sponge. Plastic fibres go down plugholes too!
- An empty plastic bottle of washing up liquid and a bar of washing up soap. Explain that the bar is so much better for the planet and that very hot water is the only vital ingredient in any washing up. Challenge people to give up the squirt!
- Cut up bits of cotton t-shirt, nighties, knickers and underpants for rags and surface cleaners. Talk about using steam/a hot cloth/old t-shirt rag instead of wipes and chemicals in plastic.
- One of those awful trigger-spray bottles! We need to avoid these altogether!

Utensils

- Wooden and metal spoons and plastic spatulas, etc.
- A stainless-steel straw and a plastic one. Talk about plastic ones getting into waterways and oceans, especially in hot countries/tourist-islands where there are a lot of bars.

- Stainless-steel cutlery. Pound and Dollar shops/stores are good for a lot of cheap natural demo-materials and some plastic ones.

Drinks

- A teabag and a tea strainer of some sort for loose tea.
- A teapot and ceramic mug would be nice on the table too.
- A coffee plunger, and a plastic coffee pod if you have one or can source from family or friends. Most coffee pods end in landfill in the trillions! You may be able to get hold of a stainless-steel one.

Grooming and hygiene

- A plastic disposable razor, and a stainless-steel one.
- A Bamboo or wooden brush with natural bristles and a rubber cushion, and a comb.
- A bar of shampoo soap.
- Some toothtabs. Lots around now.

Gardening

- Buy a couple of biodegradable fibre plant pots and show a plastic one as well. Talk about reducing the use of plastic in the garden in general.

Balloons

- 2 balloons, one whole and the other in pieces. Beg/find one, don't buy, with string attached, to demonstrate the danger to wildlife and sea life from the plastic, and strings that entangle.

- Talk about Helium being a finite source too.
- Mylar, shiny metal, balloons are even more dangerous if caught on powerlines. They explode!
- Find a model or toy turtle. Put a thin see-through plastic bag in a jar of water with a screw-lid to demonstrate how it looks like a jellyfish to a hungry turtle. Turtles have fleshy barbs in their mouths that help food go in but prevent it coming out. Simply put: Turtles can't cough it up! They are starving to death all over the planet with stomachs full of plastic!

Step 3

- Again, be as creative and you can and include some games as well as the aforementioned Fishing Game if you have room on your table/s.
- Make sea life mobiles using old cereal packet carboard – just string them onto 1 piece of garden cane at different levels.
- Create some sea life pictures for children to colour in. Turtles, dolphins, whales, sharks, seahorses, octopus and striped or spotty fish are always a draw.

Step 4

- Use a tablecloth on your table/s that will enhance the message.
- A blue one to represent the Oceans is nice.

I use a stretchy banqueting cloth. Yes, it is made of Nylon but I've had it years and it is absolutely perfect for a quick set up and take down, fixing under the legs of the table in 5 minutes. Sometimes you do need to compromise until something better for the planet comes along.

Step 5

- Make a lovely clear banner for your Day/Cause, but not a plastic one.
- You can make one out of cotton bunting. Maybe a simple 'Plastic free table/Day' one.
- Make it sturdy so it will last and can be brought out time and time again.
- Have fun, and always be creative with your designs!

Step 6

When your table is all set up and your eager audience gathers, do a demonstration of how to make shopping bags out of tank-tees. Ask around for old ones or source cheaply from charity/thrift shops.

Sewing machine method – if you can have a machine with you too. Only adults use the machine but the children can choose the tee design. They love this demo!

- Turn tee inside out and sew the bottom up.
- Turn right side out and you have a perfect bag with readymade handles.

No sew method. Anybody can do this but you will need good scissors and someone to supervise children, who love both of these bags to take home!

- Flatten out the tee.
- Cut a fringe on the bottom.
- Stretch the fringing and then tie opposite sides.
- Another perfect bag with readymade handles.

Step 7

- Do your research and talk about the local and global problem.
- Become aware of what is actually going on in the world.
- There are some exceptionally disturbing videos out there that will definitely educate us to the problem.
- Always mention The Blue Planet Effect and Sir David Attenborough. His wonderful team highlighted the horrendous situation and brought it to global attention.

Our precious Mother Earth needs us today!

'We don't need a handful of people doing zero waste perfectly. We need millions of people doing it imperfectly.' - Anne Marie Bonneau

'You cannot get through a single day without having an impact on the world around you. What you do makes a difference.' – Jane Goodall

'There is hope if we all – every single one of us – take our share of responsibility for life on Earth. Those in power can influence change.

And those with knowledge and the ability to innovate can provide solutions to a great number of problems.' - Sir David Attenborough

I wish you all the best in spreading these vital messages in your local community and beyond.

Once again, blessings and heartfelt thanks for all you do!

Recommended Organisations/ Websites and Books

The Ocean Cleanup: theoceancleanup.com
Exciting projects from a company founded by Boyan Slat.

Marine Conservation Society: mcsuk.org
The UK's leading charity for the protection of our seas, shores and wildlife.

Surfers Against Sewage: sas.org.uk
They are 'the voice of the Ocean' and 'inspire, unite and empower communities to protect oceans, beaches, waves and wildlife.'

Break free from Plastic: breakfreefromplastic.org
The Global movement to stop plastic for good.

Less Plastic: lessplastic.co.uk
You can download/buy a wide range of informative posters and postcards from their site.

The Story of Stuff Project: thestoryofstuff.org
Very clever animations about reducing plastic and waste in general.

The 5 Gyres Institute: 5gyres.org
A non-profit organisation that is working to rid the world of plastic pollution.

The Plastic Pollution Coalition: plasticpollutioncoalition.org
An American organisation fighting the plastic industry.

The Plastic Soup Foundation: plasticsoupfoundation.org
'No plastic waste in our water' is their mission.

Sea Legacy: sealegacy.org - Saving our wonderful oceans.

Seashepherd Conservation Society: seashepherd.org
An American non-profit company protecting our precious oceans.

Kids Against Plastic: kidsagainstplastic.co.uk
A charity set up by 2 young sisters to educate about plastic reduction.

Children's books

Colourful books are a good way to focus attention on global problems, and prove very popular on an Alternatives to Plastic Demo Table. Following is a small selection of the best ones but there are many more now as the world wakes up to pollution and waste.

Wild Tribe Heroes: wildtribeheroes.com
Award-winning books written by Ellie Jackson about the plight of animals versus plastic, and changes made

to animal habitats. The series, which I'm sure will be growing, is beautifully illustrated by a number of different very talented artists.

Duffy's Lucky Escape
Marli's Tangled Tale
Nelson's Dangerous Dive
Buddy's Rainforest Rescue
Hunter's Icy Adventure

Maya in the Rubbish Sea
Written by Lucy Munday and illustrated by Simona De Leo.

Saving Tally
An Adventure into the Great Pacific Plastic Patch (Save The Planet Books) – written by Serena Lane Ferrari and illustrated by Giorgia Vallicelli.

Harry Saves The Ocean
Teaching children about sea pollution and recycling, by N.G. K., Sylva Fae (Goodreads Author) and illustrated by Janelle Dimmett. (Harry The Happy Mouse Book 5).

The Lorax by Dr. Seuss
A wonderful tale of how the world can change overnight because of greed, but ultimately heal through caring. There is an animated film based on the story too.

Talking about film animation, Pixar's *Wall-e* is a marvellous Eco film that needs to be taken more seriously.

Other books:

Old Enough to Save the Planet – Contributors Anna Taylor, Adelina Lirius, Loll Kirby.

Simple Acts to Save our Planet
500 ways to make a difference, by Michelle Neff.

Plastic Game Changer by Amanda Keetley of Less Plastic. Helping businesses to go plastic free.

Salt, Lemons, Vinegar, and Baking Soda
Natural cleaning by Shea Zukowski.

There are also some great flashcards for teaching children available online.

Eco Books: ecobooks.com – books on Ecology and the Environment.

Magazines:

Pebble Mag: pebblemag.com
Has all the latest enviro-news and Earth-friendly buys.

Eco Kids Planet: ecokidsplanet.co.uk
The best Science and Nature magazine for children aged 7 – 11.

Businesses

These are some of my favourite go-to sites/shops and online businesses for plastic free buys/saving the planet inspiration.

Support your local zero waste/plastic free stores when you can.

The following businesses are correct as of 2021, and are not in alphabetical order:

Theveganandecostore.co.uk is a small and passionate business in Dover, Kent.

Thethoughtfulpup.co.uk stocks natural leads, collars and accessories.

Rapanuiclothing.com does all-natural clothing.

Gumbies.co.uk/com, an Australian company, has flip flops in recycled rubber, recycled textile and natural fibres.

Wavesflipflops.co.uk sell 100% natural rubber flip flops.

Nonplasticshop.co.uk stocks plastic free goodies.

Babipur.co.uk – 'Ethical shopping for kids.'

Andkeep.com sells a huge range of plastic free items.

Cariuma.com does ethically made sneakers from Rio de Janeiro, Brazil.

Handmadenaturals.co.uk has hand creams/potions and lotions.

Plasticfreepantry.co.uk stocks among other things, cereals, pulses, their own plastic-free spaghetti and coffee.

Cosycottagesoap.co.uk has lovely affordable hand and body soaps, shampoo bars, and washing-up bar soap. You can build a bundle. Get together with family/ friends and save money!

Woolovers.com sells 100% wool clothing, accessories, blankets and shawls at affordable prices.

Ecolunchbox.com is an American company that does great stainless-steel lunchboxes, and has a few UK stockists such as Babi Pur.

Biome.com is an Australian company that sells eco-friendly/plastic-free products.

Wearewild.com
'Wild is a sustainable Natural Deodorant delivered straight to your door. Aluminium free with compostable plastic free refills and a 100% effective formula.'

Hunter.com sells real rubber boots.

Seedandbean.co.uk has plastic-free chocolate. The outer paper layer of their wrappers is recyclable and the inner foil is Natureflex™, made from eucalyptus wood pulp, and fully compostable.

Playinchoc.com do organic chocolates + 3D puzzles and fun facts cards to play with, learn about and collect. Plastic free too!

Ecoenfys.co.uk makes environmentally-friendly play dough. Handmade in Wales.

Milksafes.co.uk make boxes to protect glass milk-bottles from the weather, birds and thieves. Several designs available as well as a bespoke service.

Greenfibres.com sell organic and ethically-made bedding and clothing.

Baavet.co.uk specialises in wool duvets, mattress-toppers and pillows.

Envirotoy.co.uk sell plastic free eco beach and water toys.

Twofarmers.co.uk make crisps with their own Herefordshire potatoes in plastic free packaging that is 100% home compostable. They also do sharing-tins.

Zealoptics.com make 100% bioplastic sunglasses.

Perfectpartyboxes.co.uk offer themed party-packs with no plastic tat.

Ecofriendlyshop.co.uk sells lots of plastic free goodies.

Milly&sissy.co.uk do dried powdered soap sent out in compostable pouches.

Adropintheocean.com sells reef-safe sunscreen that comes in a refillable bottle.

- When the aluminium bottle is empty you go online and order a refill.
- They'll post a new aluminium bottle of sunscreen to you along with a prepaid label for your old bottle, which they then sanitise and reuse.

Perfect circular economy.

Beautykitchen.co.uk - sustainable, cruelty free, and certified microplastic free. Their natural hand sanitiser is in a choice of two sizes of large refillable aluminium bottles. These can then be returned for cleaning and reuse. They come with a small handy aluminium or glass spray-bottle. There is an optional sprayer for the larger ones too so you can keep a family-sized one handy. They too have the perfect circular economy for all their products:

- Return
- Refill
- Repeat

Naturiolcollective.co.uk sells ethical and plastic free goodies. Based in Conwy, North Wales.

Allbirds.co.uk do sustainable footwear made of natural materials including Merino wool and Eucalyptus.

Ecokindly.co.uk.
Their range is natural, sustainable, organic, Vegan, cruelty free, and has zero waste packaging.

Friendlysoap.co.uk make soaps and shampoo bars with all the frees we need, at affordable prices.

Theplasticfreeplanet.co.uk is run by a mother and daughter team helping to save Mother Earth.

Bambooi.co.uk has lots of lovely affordable Bamboo products. They have a 'build a bundle' option. Get together with family/friends and save money.

Loamandlore.com do biodegradable and compostable phone-cases.

Ropelocker.co.uk sells 100% cotton rope for house and garden.

Greengolfuk.co.uk is a Yorkshire-based golf goods retailer. Their products are specially designed and manufactured so they cause as little harm to the environment as possible.

Hangersnow.co.uk sell sustainable hangers made from wood, bamboo, recycled card, and wood pulp

Kiltane.com/harris-tweed-dog-lead. Not entirely plastic free but it's better than all plastic, which most leads are if they arc not leather.

Smilersmdf.co.uk make a huge variety of inexpensive MDF blanks for crafting. Lots of designs to paint/customise, a lovely alternative way to decorate any occasion plastic free. They even do bunting. There is a bespoke service for some of the designs.

Lightning Source UK Ltd.
Milton Keynes UK
UKHW010910141221
395640UK00003B/392

9 781839 757662